A BASIC/INTERMEDIATE COURSE
FOR WATER SYSTEM OPERATORS

Volume 1
INTRODUCTION TO

# Water Sources
# and
# Transmission

PRINCIPLES and PRACTICES of
WATER SUPPLY OPERATIONS

American Water Works Association
6666 W. Quincy Ave., Denver, Colorado 80235

ISBN 0-89867-177-9

# Foreword

*Introduction to Water Sources and Transmission* is the first volume in a five-part series designed for use in a comprehensive training program titled "Principles and Practices of Water Supply Operations." Volume 1 contains background information on water sources and characteristics, factors influencing water demand, the development of water sources, and transmission systems.

Other student volumes in this series include:

| | |
|---|---|
| Volume 2 | *Introduction to Water Treatment* |
| Volume 3 | *Introduction to Water Distribution* |
| Volume 4 | *Introduction to Water Quality Analyses* |
| Reference Handbook | *Basic Science Concepts and Applications* |

In addition, an instructor guide and solutions manual is available for use with volumes 1 through 4, and audiovisual presentations have been prepared to supplement volumes 2 and 3.

Course content of this series has been specifically designed to meet the training requirements of basic- to intermediate-grade water treatment and water distribution operators. A modular format has been developed throughout the series to provide flexibility in conducting short or long-term training courses.

The reference handbook, a companion volume correlated and cross-referenced with volumes 1 through 4, is a complete student resource of background reading and applied operational problems in the areas of basic mathematics, hydraulics, and chemistry.

It is the hope of the American Water Works Association that the material contained in this series will provide a comprehensive instructional delivery system for the training of water supply operators. This training project was made possible by funding from the US Environmental Protection Agency, Office of Drinking Water, under Grant Number T900632-01 awarded to the American Water Works Association.

## Disclaimer

Several photographs and illustrative drawings that appear in this volume have been furnished through the courtesy of various product distributors and manufacturers. Any mention of trade names, commercial products, or services does not constitute endorsement or recommendation for use by the American Water Works Association or the US Environmental Protection Agency.

# Acknowledgments

Publication of this volume was made possible through a grant from the US Environmental Protection Agency, Office of Drinking Water, under Grant No. T900632-01 in recognition of the need for a comprehensive instructional approach to operator training in the water supply industry. John B. Mannion, special assistant for communications and training, represented the Environmental Protection Agency, Office of Drinking Water, as project officer, and Bill D. Haskins, director of education, served as project manager for the American Water Works Association.

Joanne Kirkpatrick and Benton C. Price, under contract with VTN Colorado, Inc., were the principal developers for this volume of work as well as the instructor guide and solutions manual.

Special thanks is extended to the many individuals who gave liberally of their time and expertise in the technical review of all or portions of the manuscript. The following are credited for their participation on the review committee or as an independent reviewer:

E. Elwin Arasmith, Instructor, Water/Wastewater Technology, Linn-Benton Community College, Albany, OR

Edward W. Bailey, Manager, Water Division, Colorado Springs, CO

James O. Bryant Jr., Director, Environmental Resources Training Center, Southern Illinois University at Edwardsville

Earle Eagle, Staff Coordinator, South Carolina Board for Technical & Comprehensive Education

Clifford H. Fore, Associate Director, Environmental Resources Training Center, Southern Illinois University at Edwardsville

James T. Harvey, Superintendent of Production, Water Works, Little Rock, AR

William R. Hill, Director of Technical Services, Floyd G. Browne & Associates, Ltd., Marion, OH

Jack W. Hoffbuhr, Chief, Water Supply Section, US Environmental Protection Agency, Region VIII

Kenneth D. Kerri, Professor of Civil Engineering, School of Engineering, California State University at Sacramento

Jack E. Layne, Deputy Director of Engineering & Construction, Water Department, Denver, CO

Ralph W. Leidholdt, Water Supply Specialist, Water Quality Control Division, Colorado Department of Health

L.H. Lockhart, Coordinator, Manpower Planning, Development & Training, Administrative Division, South Carolina Department of Health & Environmental Control

Andrew J. Piatek Jr., Chief Operator & Superintendent, Borough of Sayreville, NJ

Fred H. Soland, Engineer, Water Supply Division, South Carolina Department of Health & Environmental Control

Robert K. Weir, Design Engineer, Foothills Project, Water Department, Denver, CO

Glenn A. Wilson, Superintendent, Source of Supply Section, Plant Division, Water Department, Denver, CO

Leonard E. Wrigley, Staff Director, South Carolina Water Association

Robert L. Wubbena, Vice President, Economic and Engineering Services, Inc., Olympia, WA

# Table of Contents

# Water Sources and Transmission

# Module 1

# Sources and Characteristics

One of the first things a water treatment plant operator should learn is where the water entering the plant comes from. This is important because the water source affects both the *quantity* and *quality* of raw water available for treatment.

It is especially important to understand how water acquires its various characteristics. Understanding these characteristics and their effect on safe, good-tasting drinking water is a necessary first step toward understanding water treatment.

This module contains discussions of how water is formed and then collected, what some of the characteristics of water are, how these characteristics come about, and how water quality characteristics affect the safety and PALATABILITY* of water.

After studying this module, you should be able to

- Define the water cycle and define the major elements that make up the water cycle.

- Define surface water; list some of the factors controlling runoff; define and be able to sketch a watershed; discuss the purpose of water laws.

- Define ground water; describe the difference between a water table aquifer and an artesian aquifer and be able to sketch both.

- Identify the physical, chemical, biological, and radiological characteristics of water and discuss the significance of each. Analyze water quality operating records. Describe some of the ways in which surface water acquires its characteristics. Describe some of the ways in which ground water acquires its characteristics.

- List the characteristics of water that have public health significance. Define the term "maximum contaminant level." Identify typical waterborne microbiological organisms and the diseases that they cause.

---

*Words set in SMALL CAPITAL LETTERS are glossary terms. Definitions for these terms can be found in the glossary at the end of this volume.

## 1-1. The Water Cycle—*The Fresh Water Delivery System*

The constant movement of water from the earth to the clouds and back again is called the water cycle. Most aspects of this cycle are as obvious to you as the fact that water runs downhill. However, the following brief discussion of the cycle will introduce the correct terminology. Some terms, such as condensation, infiltration, percolation, and transpiration, are not commonly understood.

The water cycle may be described as four events which occur over and over again, as shown in Figure 1.

### Water Moves From Earth to Sky: *Evaporation and Transpiration*

One way that water moves from the earth to the sky is by EVAPORATION. Heat from the sun evaporates water from the ocean and other bodies of water, such as rivers, lakes, and streams. The water vapor rises into the air and is held there by moving air masses. This can be thought of as "cloud storage."

TRANSPIRATION is the other way that water moves from the earth to the sky. In this process, water from the soil is taken up by the roots of trees, shrubs, grasses, and other plants, and is given off to the air through tiny holes in the underside of the leaves.

### Water Vapor Forms Tiny Water Droplets: *Condensation*

Under certain conditions the water vapor stored in clouds begins to form tiny droplets of water or minute ice crystals. This process is called CONDENSATION. The condensed water appears as dark storm clouds.

### Water Falls Back to Earth: *Precipitation*

As the water droplets and crystals gather together and become heavy enough, they fall back to the earth as PRECIPITATION. The world's entire fresh water supply is delivered by precipitation, most often in the form of rain, hail, sleet, or snow.

### Water on the Earth Penetrates The Ground or Runs off the Surface: *Infiltration, Percolation, Surface Runoff*

After wetting the foliage and ground, part of the precipitation soaks into the soil—a process called INFILTRATION. Much of the water infiltrating the soil is held in the root zone, which is the surface layer of soil ranging in depth from several inches to several feet. Some of this water is taken up by the roots of plants and returned to the air by transpiration.

A portion of the infiltrated water is drawn back to the surface by the CAPILLARY ACTION of the soil. At the surface, solar heat evaporates the water and returns it to the air.

The remaining infiltrated water continues to move downward below the root zone until it enters the ground-water AQUIFER. This deep vertical movement of water is called PERCOLATION. The GROUND WATER moves through the pores of the aquifer materials and may reappear at the surface in areas at lower elevation than where it entered the aquifer. At such places, ground water discharges

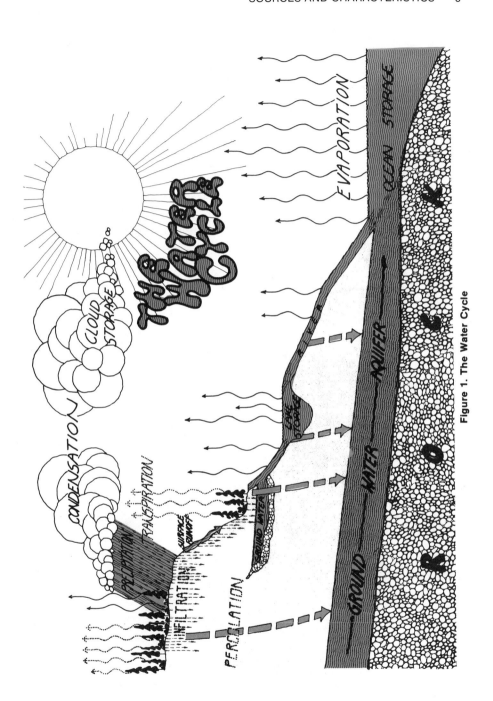

Figure 1. The Water Cycle

naturally in the form of springs and seeps, which sometimes maintain the flow of streams in dry periods. The streams, carrying both surface runoff and natural ground-water seepage, eventually lead back to the oceans.

When the surface soils can no longer take in water—a condition called SATURATION—excess precipitation begins to flow downhill over the surface. This water, called SURFACE RUNOFF, moves over the land, into rivulets, gulleys, creeks, streams, rivers, and lakes, and eventually returns to the ocean. This completes the water cycle.

## 1-2. Surface Water—*Water on the Land*

SURFACE WATER is the major source for public water supply systems. This is because large cities draw most of their water from surface reservoirs. About three out of every four people in the US drink water that originated from surface water sources, so it is important for you to have a thorough understanding of the factors that influence surface water flows.

### Origin of Surface Waters

Surface waters come from two sources:

- Precipitation

- Ground water

As discussed briefly in the preceding section, when rainfall reaches the ground it either infiltrates into the soil, evaporates into the air, or runs off as surface water. But rainfall is not the only form of precipitation that results in surface water runoff. Snow, which may remain on the ground for many months, eventually melts and also contributes to surface water runoff. In portions of the western US, melting snow produces the major part of the annual runoff.

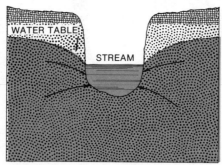

**Figure 2. Surface Flow From Ground-Water Seepage**

If rainfall were the only source of surface waters, then all streams and rivers would dry up shortly after a rain; however, many streams and rivers flow throughout the year. This is due in part to snowmelt and in part to ground water that enters streams and rivers from springs and seeps. Figure 2 shows an example of a stream that is receiving surface flow from ground-water seepage.

Just as ground water can give up water to a stream, streams give up a portion of their flows to RECHARGE ground waters, as shown in figures 3A and 3B.

**A.**

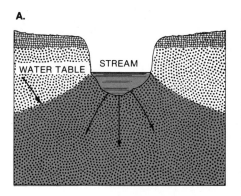

**B.**

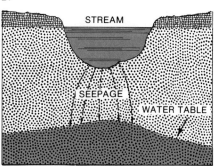

**Figure 3. Stream Flow Recharging Ground Water**
A. High Water Table     B. Low Water Table
Reprinted with permission of Johnson Division, UOP Inc.,
Saint Paul, Minnesota, from *Ground Water and Wells*, copyright ©1975

## Factors That Control Surface Water Runoff

There are a variety of factors that affect surface runoff. Perhaps the most significant are

- Rainfall intensity

- Duration of rainfall

- Soil composition

- Soil moisture

- Slope of the ground

- Vegetation covering the ground

- Man-made influences

Slow, gentle rainfalls usually produce very little runoff. There is plenty of time for the rain to soak into (infiltrate) the soil. However, as RAINFALL INTENSITY increases (measured in inches per hour), the surface of the soil becomes saturated. Since a saturated soil can hold no more water, further rainfall builds up on the surface and begins to run off, creating surface water flow.

The *duration of the rain storm* also influences the amount of runoff. Even a slow, gentle rain, if it lasts long enough, can eventually saturate the soil and produce runoff.

The *soil composition* has a marked effect on the amount of runoff produced. Beach sand, for example, has large VOID SPACES and therefore allows water to pass through readily. A high intensity rainfall, approximately an inch of rain an hour, on such a soil may result in little, if any, runoff. Yet even a low intensity rainfall of 0.1 inches per hour can result in a substantial runoff on a soil that is tightly packed, such as clay. Clay soils not only have small void spaces, but they swell when wet. This expansion closes the void spaces, which further reduces the infiltration rate and results in greater surface runoff.

If soil is already wet from a previous rain, surface runoff will occur sooner than if the soil were dry. Consequently, existing *soil moisture* has an effect on surface runoff.

Surface runoff is affected by the *slope of the ground.* In general, flat ground only allows 20 percent of the rain to run off. However, on steeply sloping ground, as much as 80 percent of the rain becomes surface runoff.

The amount of *vegetation covering the ground* plays an important role in runoff. Well-planted soils produce less runoff than bare soils. Roots of plants and deposits of decaying natural organic litter (such as leaves, branches, or pine needles) increase the void spaces in the soil and allow water to move easily into the soil.

Hard driving rains will sometimes compact a soil, closing the void spaces. However, vegetation and the organic litter act as a cover to protect soil from this compaction, and to help maintain the soil's water-holding and infiltration capacity. This cover also reduces the amount of evaporation of soil moisture.

*Man-made influences* have a decided effect on surface water runoff. Dams control it; channels, canals, and ditches divert it; and streets and other paved areas increase it.

Perhaps you can begin to see how these various factors become interrelated. For example, a sloped terrain will not produce more surface runoff than a flat terrain if other factors prevail; nor will a plot that is bare of vegetation necessarily result in greater surface runoff than a vegetated plot.


## How Surface Water Is Collected

After surface water runoff has been produced, it flows in the path of least resistance. It begins to form rivulets, which will often then flow into brooks, creeks, and rivers. Each rivulet, brook, creek, and stream receives water from an area of land surface that slopes down toward one primary watercourse. This drainage area is known as a WATERSHED or DRAINAGE BASIN. As shown in Figure 4, a watershed is a basin surrounded by a ridge of high ground. The ridge, called the WATERSHED DIVIDE, separates one watershed from another.

Within a watershed there are two types of surface waters: (1) WATERCOURSES, and (2) WATER BODIES. Watercourses convey surface waters from higher

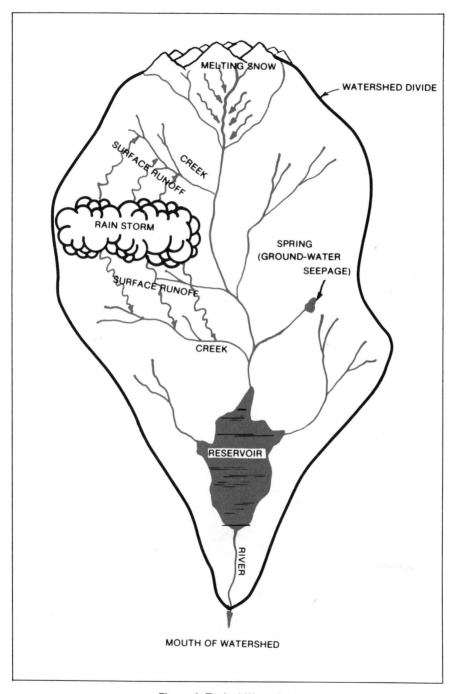

**Figure 4. Typical Watershed**

elevations to lower elevations. Typical natural and man-made watercourses include:

| Natural | Man-Made |
|---------|----------|
| Brooks | Ditches |
| Creeks | Channels |
| Streams | Canals |
| Rivers | Aqueducts |

Natural watercourses may flow continuously or only occasionally. Continuously flowing streams are called PERENNIAL streams. These streams are supplied both by surface runoff and by springs, and ground-water seepage. Streams flowing only occasionally are called EPHEMERAL streams. Ephemeral streams usually flow only during and shortly after a rain, and are supplied only by surface runoff.

Man-made watercourses carry water only when man intentionally diverts water to them.

A water body is a water-storage basin. It can be a natural basin such as a pond or lake, or a man-made basin such as a reservoir. Man-made water bodies are built to serve some specific water need. These bodies range from basins for watering stock to massive reservoirs that store water for municipal and recreational use. Figures 5 and 6 show examples of natural and man-made water bodies.

**Figure 5. A Natural Lake (Lake Mary, California)**

*Courtesy Delaware State Development Dept., Dover DE*

**Figure 6. A Man-Made Reservoir (Hoops Reservoir, Delaware)**

## Water Rights

Water is a valuable commodity everywhere, but particularly in regions where water is in short supply. A system of laws has been developed to establish who has the right to use what waters. Water rights play an important role in determining the availability of water in many parts of the world. Water law and water rights are highly complex subjects. This brief discussion is designed to provide you with an overview of two major types of water rights, and of how they have been applied in the US.

**Riparian rights doctrine.** Under the original concept of RIPARIAN RIGHTS, the owner of land next to a stream (riparian land) is entitled to receive the full natural flow of the stream unchanged in quality or quantity, and he is therefore protected against the diversion of water upstream from his property. He is also protected from the diversion of excess floodwaters toward his property. In other words, no upstream owner may materially lessen or increase the natural flow of a stream to the disadvantage of a downstream owner. The water must be used on the riparian land and it cannot be transported out of the watershed. It can only be used for domestic purposes and for watering domestic stock—these are called *ordinary* or *natural* uses.

The original English riparian doctrine has a serious defect for modern society: it does not provide for use of water by the riparian owners for irrigation, nor for the sale or use of water to benefit non-riparian land. Consequently, the riparian concept was modified in the US to permit *reasonable use* of water. Reasonable use allows riparian owners to divert and use streamflow in *reasonable amounts* for *beneficial purposes*. In regions of ample flow this permits riparian owners to use all the water they need, but if the flow is inadequate for all owners, then the available water must be divided on some equitable basis. This equitable basis must allow an upstream riparian user to take as much water as he needs for ordinary or natural uses.

In general, water laws of states in the eastern part of the US follow the riparian doctrine. Riparian rights are best suited for areas that, like the East, have abundant supplies of water located where they can conveniently meet the needs for water.

However, the arid West does not enjoy that same abundant supply of water. Water users may be 500 to 1000 miles away from their water supply, which may be located in another state. The riparian laws would make it unreasonably difficult for distant users to gain the right to use such remote water supplies. More importantly, a remote source could not be relied on, since in time of water shortage the distant user would be the last to have legal claim to the water.

**Appropriative rights doctrine.** The APPROPRIATIVE RIGHTS doctrine provides for acquiring rights to use water by *diverting* it and putting it to *beneficial use.* Procedures for this are established by state statutes and courts. Water to which rights are thus acquired, called appropriated water, may be used on lands away from the stream, as well as on lands adjoining the stream. The most important feature of the doctrine of appropriation is the concept, "first in time, first in right." This means that the right of the first (or earliest) appropriator is superior to any other claim, and further appropriation is possible only if water in excess of earlier claims is available. During water shortages the available supply is *not apportioned* among all users. Instead, those claimants with the earliest priority are entitled to their full share, and if they take all the water, then those with later priorities must do without. Seventeen western states determine water rights mainly by the appropriation doctrine. These states have complex statutory water codes that govern the acquisition and control of water rights.

## 1-3. Ground Water—*The Hidden Resource*

Ground water, withdrawn from wells, is the source of about 20 percent of the water used in the US. Ground-water sources are often relatively simple to develop, and ground water usually has fewer contaminants than surface water. Small systems often find ground water the most economical water source.

### What Is Ground Water

Ground water occurs when water percolates down to the water table through the void spaces and the cracks in soil and rock. An easy way to visualize ground water is to fill a glass bowl halfway with sand, as shown in Figure 7. If water is poured onto the sand, it moves through the sand and seeps down through the voids until it comes to the watertight (impermeable) bottom of the bowl. As more water is added, the sand becomes SATURATED and the water surface (below the top of the sand) rises. The water in this saturated sand is called GROUND WATER. The sand that holds the ground water is called an AQUIFER.

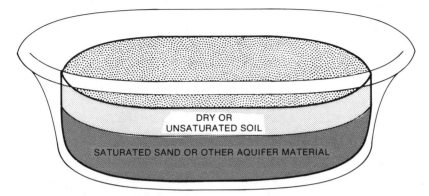

Figure 7. Example of Ground Water

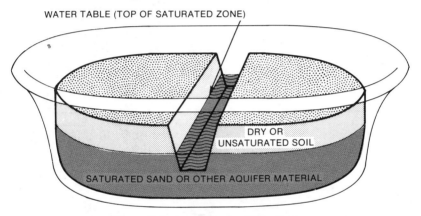

Figure 8. Determining the Water Table

When there is enough water in the bowl to saturate some of the sand, as in Figure 8, the level of the water surface can be found by poking a hole or cutting a "stream channel" in the sand. As shown in Figure 8, the level of water in the channel is the same as the level of the water surface throughout the sand. This level is called the WATER TABLE. The water table is the *top* of the ground water, or top of the zone of saturation.

## Sources of Ground Water

Half of the world's ground-water resource is located within one-half mile of the ground surface. The other half can be found in deep aquifers.

The aquifer is generally an underground layer of gravel, sand, sandstone, shattered rock, or limestone. And underneath, forming the bottom of the aquifer, is an IMPERMEABLE LAYER of rock or mineral, such as clay or granite, that keeps the water from sinking further downward.

There can be more than one aquifer in a particular area, such as where there is a layer of gravel, a layer of granite, another layer of gravel, and another impermeable layer.

Aquifers are classified as either water table aquifers or artesian aquifers, according to whether the upper water surface is unconfined or confined. A WATER TABLE AQUIFER, shown in the left hand portion of Figure 9, is one that is confined only by a lower impermeable layer. Wells constructed in those aquifers are called WATER TABLE WELLS. Water in these wells will rise only as high as the water table.

An ARTESIAN AQUIFER is one in which the water is confined by both an upper and a lower impermeable layer. The water in this "aquifer sandwich" is usually under pressure; therefore, when a well is drilled into the upper confining layer, the pressure in the aquifer forces the water to rise up above the confining layer. As shown in the center and left of Figure 9, sometimes the water will rise above the ground surface (flowing ARTESIAN WELL), and sometimes it will not rise

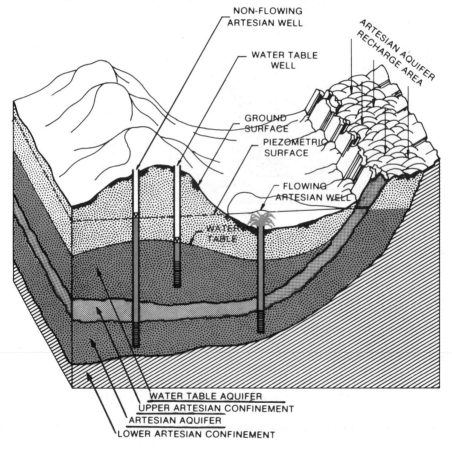

Figure 9. Water Table and Artesian Aquifers

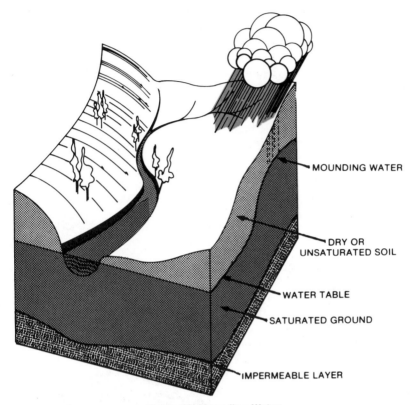

**Figure 10. Mounting Water**

as high as the ground surface (non-flowing ARTESIAN WELL). The height to which water will rise in wells located throughout an artesian aquifer is called the PIEZOMETRIC SURFACE.[1] This surface may be above the ground surface at some points and below the ground surface at other points of the same aquifer, as shown in Figure 9.

## Movement of Ground Water

Water naturally moves downhill toward the lowest point. In the examples shown in figures 7 and 8, the water will not move anywhere because the water table is flat; however, water tables are usually not flat—figures 10 and 11 show how water tables actually occur in nature. When rain falls on the watershed (Figure 10), part of the rain percolates downward to the water table, and a mound of water within the aquifer is built up above the level of the rest of the water table. The water within this mound slowly flows downslope, increasing the

---

[1] *Basic Science Concepts and Applications,* Hydraulics Section, Piezometric Surface and Hydraulic Gradeline.

level of the water table slightly and, if the water table is high enough, draining into the stream channel. In most areas, more rain occurs before the mound completely drains off, and the water table never becomes level.

Figure 11 is illustrated as though the ground surface had been lifted to show the movement of ground water. Notice that the water table is continuous and sloped, and that the ground water is moving downhill toward the lowest point— the stream channel.

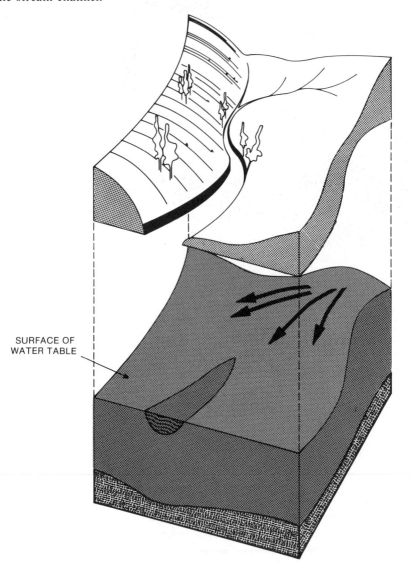

SURFACE OF
WATER TABLE

**Figure 11. Ground-Water Movement**

# 1-4.  Characteristics of Water

Water quality characteristics fall into four broad categories:

- Physical
- Chemical
- Biological
- Radiological

The quality of water is determined to a great extent by measuring the concentration of the impurities in it. The smallest unit of water, the water MOLECULE ($H_2O$), contains two ATOMS of hydrogen (H) and one ATOM of oxygen (O), and anything that is found in the water aside from water molecules may be considered as an impurity.[2] In the natural environment there is no such thing as pure water. All water contains impurities. Some impurities, such as fluoride or calcium, are beneficial; others, like arsenic or disease-causing bacteria, can be detrimental. The concentration of impurities is usually so small that it is measured in MILLIGRAMS PER LITRE (mg/L). This means that the impurities in a standard volume (a litre) of water are measured by weight (in milligrams). A concentration of 1 mg/L of magnesium in water would be about the same as 0.00013 ounces of magnesium in each gallon of water. In water, concentrations expressed as a certain number of milligrams per litre, in the range of 0–2000 mg/L, are roughly equivalent to concentrations expressed as the same number of PARTS PER MILLION (ppm). For example, "12 mg/L of calcium in water" expresses roughly the same concentration as "12 ppm calcium in water." However, mg/L is the measurement preferred over ppm.[3]

**Table 1.  Typical Surface-Water Mineral Concentrations**

| Source of Water | Total Dissolved Minerals (mg/L) |
|---|---:|
| Distilled | 0 |
| Rain | 10 |
| Lake Tahoe | 70 |
| Suwannee River | 150 |
| Lake Michigan | 170 |
| Missouri River | 360 |
| Pecos River | 2,600 |
| Ocean | 35,000 |
| Brine well | 125,000 |
| Dead Sea | 250,000 |

The examples given in Table 1 should help you develop a feeling for the relative magnitudes of mineral concentrations in water.

---

[2]*Basic Science Concepts and Applications,* Chemistry Section, The Nature of Matter.
[3]*Basic Science Concepts and Applications,* Chemistry Section, Concentrations.

River water usually has a dissolved-minerals concentration of less than 500 mg/L, although as shown in Table 1, some river waters may contain concentrations of 2000 mg/L, or more. Mineral concentrations in ground water can vary from several hundred mg/L to more than 10,000 mg/L. The mineral content of any water consists of individual chemicals, such as calcium, magnesium, sodium, iron, and manganese.

## Physical Characteristics

The physical properties of water include temperature, turbidity, color, and taste and odor. A brief discussion of these properties follows. A more detailed presentation for each test can be found in Module 5 of Volume 4, *Introduction to Water Quality Analyses.*

**Temperature.** Water temperature is important in water treatment operations. For example, chemicals used in water treatment dissolve more easily in warm water than in cold water. Particles will settle out more quickly in warm water. Warmer temperatures also encourage the growth of various forms of plant life in water. Yet, from the consumer's standpoint cold water is preferred because it tastes better and, when used as a coolant in air conditioners or industry, it is more efficient.

In reservoirs and lakes, water temperatures are partly the cause of seasonal turnovers. Temperature in lakes and reservoirs varies with depth. Normally the water is colder (and heavier) at the bottom, and warmer (and lighter) near the top. Twice a year there can be a change—the surface water becomes slightly heavier than the water below. Encouraged by the mixing action of wind, the water in the lake turns over. The turbulence created by this TURNOVER causes the bottom water to carry decomposed bottom deposits to the surface. This creates a very unpleasant odor in the water and can cause quality and treatment problems for several days or weeks. (Turnover is discussed more fully later in this module.)

Knowing the water temperature at various levels in a reservoir can help the operator anticipate turnover and select the level of withdrawal that will produce the best quality water.

TEMPERATURE is measured on either of two scales: (1) the Fahrenheit (F) scale or (2) the Celsius (C) scale. The freezing point of water is 32°F, or 0°C; the boiling point is 212°F, or 100°C.[4]

**Turbidity.** The cloudiness of water is called TURBIDITY. When a beam of light passes through cloudy water some of the light reflects off the particles suspended in the water. The amount of light reflected by the particles is a measure of the water's turbidity, expressed in TURBIDITY UNITS (TU).

The particles causing turbidity are fine insoluble matter, either INORGANIC MATERIAL such as clay, silt, or sand, or ORGANIC MATERIAL such as algae and leaf particles. Industrial and domestic wastes contribute both inorganic and organic materials. Practically all public water supplies that are filtered are free from noticeable turbidity.

---

[4] *Basic Science Concepts and Applications,* Mathematics Section, Conversions (Temperature).

Because of the natural filtering effect through soils that ground water receives, ground-water turbidity is often near to zero. Surface water turbidities vary widely, from less than one TU to as high as 200 TU or more.

**Color.** The physical characteristic COLOR occurs primarily in surface waters. It usually indicates the presence of decomposed organic material or certain inorganics such as iron and manganese. The organic material can come from either natural sources, such as leaves, roots, or plant remains, or from man-made sources, such as domestic or industrial wastewaters.

Color is measured by comparing the color of a sample with the color of a standard chemical solution. The units of measure are COLOR UNITS, abbreviated CU. Usually a color of less than 15 CU passes unnoticed, while a color of 100 CU has the appearance of a light tea. Figure 12 shows a colorimeter—the device used to measure color.

Highly colored water is objectionable for most industrial uses, and from an AESTHETIC standpoint, it is unsuitable for drinking water. Color in water can also indicate POLLUTION of the water by natural or man-made inorganics or organics, perhaps making the water unsafe to drink.

**Taste and odor.** Taste and odor in water can be caused by a wide variety of materials, such as ALGAE or other microorganisms, decaying organic matter,

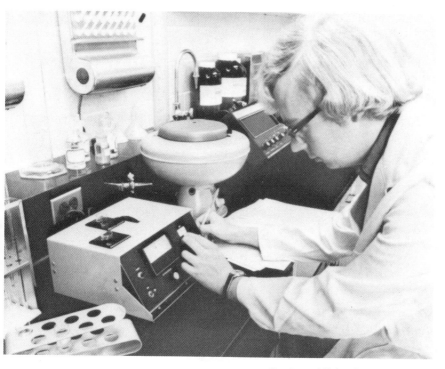

*Courtesy of Fisher Scientific Company*

**Figure 12. Color Tests in Treatment Plant Lab**

wastewater and industrial wastes, minerals, and dissolved gases such as hydrogen sulfide or chlorine. Highly mineralized waters have a medicinal or metallic taste. Distilled water, on the other hand, tastes flat because it lacks the minerals and dissolved gases that, in small amounts, can give water a pleasant taste. Both taste and odor become more noticeable as the water temperature increases.

Although the senses of taste and smell are closely related, the sense of smell is far more discriminating than the sense of taste. Therefore, in water treatment we usually measure only odor. Odors are measured by persons who smell a series of diluted water samples. The test begins by smelling a sample with no odor (pure distilled water) and continues by smelling samples containing less and less distilled water and more and more sample, until an odor is detected. The dilution at which an odor is first detected is called the THRESHOLD ODOR. The particular dilution of that sample is called the threshold odor number (TON).

## Chemical Characteristics

Chemical characteristics of water fall into two categories: (1) inorganic characteristics, and (2) organic characteristics.

*Inorganic characteristics* include pH, hardness, dissolved oxygen, dissolved solids, and electrical conductivity.

**pH.** The term pH is used to express the acid or alkaline condition of a solution. The pH scale runs from 0 to 14. On this scale, 7 is neutral; 7 is the pH of pure water. A pH less than 7 indicates an ACIDIC WATER. A pH greater than 7 indicates a BASIC or ALKALINE WATER. The normal range of surface water pH is 6.5 to 8.5; the pH of ground water ranges from 6.0 to 8.5.

A typical pH meter is shown in operation in Figure 13.

*Courtesy of Fisher Scientific Company*

**Figure 13. pH Testing in the Field**

**Figure 14. Corrosion**

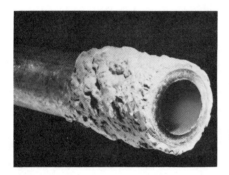 

*Courtesy of Johnson Controls, Inc.*

**Figure 15. Typical Pipe Scaling Problem**

The pH of water can have a marked effect on treatment plant equipment and processes. At a pH value less than 7, water tends to CORRODE the equipment and other materials that it contacts (Figure 14).

At a pH value greater than 7, water has a tendency to deposit scale, which is particularly noticeable in pipelines and hot-water home appliances (Figure 15).

**Hardness.** A "hard" water is any water that contains significant amounts of calcium and magnesium. Hard water can be a problem in water supply and treatment because, like water with a high pH, it can cause scale to form in pipes and meters, which reduces pipe capacity and can cause meters to malfunction. If the water is hot, as in boilers and hot-water lines, the scale forms much faster. One millimetre of deposited scale can increase the cost of heating hot water by more than 10 percent. Hard water may have an objectionable taste and may require more soap when used for washing. A standard laboratory test is used to determine HARDNESS. The results are expressed in mg/L as calcium carbonate[5] ($CaCO_3$).

---

[5] *Basic Science Concepts and Applications,* Chemistry Section, Solutions (Normality).

The following table summarizes typical ranges of hardness:

**Hardness Concentration**
*mg/L as CaCO₃*

| | |
|---|---|
| 0–60 | Soft |
| 61–120 | Moderately hard |
| 121–180 | Hard |
| More than 180 | Very hard |

**Dissolved oxygen.** One of the more common dissolved gases in water is oxygen. Although DISSOLVED OXYGEN (DO) is absolutely vital for the support of fish and other AQUATIC LIFE, it can also cause corrosion.

Oxygen can become dissolved in water in three ways:

1. The oxygen enters the water directly from the air (natural AERATION).

2. Oxygen is introduced into the water by algae (PHOTOSYNTHESIS).

3. Oxygen is introduced by mechanical equipment (mechanical or diffused AERATION).

In natural waters DO is almost always present to some extent. A warm-water lake will usually contain about 5 mg/L of DO; a cold-water lake will usually contain about 7 mg/L or more. However, if algae are present in the water, the DO level will vary greatly over a period of a day. During sunlight hours, algae

*Courtesy of Fisher Scientific Company*

**Figure 16. Dissolved Oxygen Meter in Use**

produce oxygen so that the DO level rises. During the night, algae use the oxygen, which causes the DO level to drop.

Although DO is a healthy and necessary condition in water, it is also the most important cause of corrosion. Corrosion can occur whenever metal, water, and oxygen come into contact. Consequently, water containing DO will attack the metallic surfaces that it contacts—such as pipe, meters, pumps, and boilers. Generally, the higher the DO concentration is, the more rapid the corrosion will be.

DO can be measured either chemically or electrically. The measured concentration of DO is reported in milligrams per litre. A typical DO meter is shown in Figure 16.

**Dissolved solids.** Because of the remarkable dissolving properties of water, surface waters and ground waters dissolve minerals from the soil and rock materials that they contact. Table 2 gives a comparison of the chemical quality of four water supplies.

Water can contain such toxic dissolved minerals as arsenic, barium, cadmium, chromium, lead, mercury, selenium, and silver. These minerals and their health effects are discussed under "Public Health Significance," later in this module.

The DISSOLVED SOLIDS content of water is a general indicator of its acceptability for drinking, agricultural, and industrial uses. Waters high in dissolved solids can create problems such as taste, odor, hardness, corrosion, and scaling. Daily users as well as occasional users of such water may experience laxative effects. For these reasons a dissolved solids limit of 500 mg/L is recommended for

**Table 2.  Chemical Quality Comparison**

| Chemical | Analyses of Water[1] mg/L | | | |
|---|---|---|---|---|
| | River[2] | Well[3] | Canal[4] | Lake[5] |
| Silica ($SiO_2$) | 5.4 | 41. | 6.6 | 11. |
| Iron (Fe) | 0.11 | 0.04 | 0.11 | 0.10 |
| Calcium (Ca) | 9.6 | 50. | 83. | 2.9 |
| Magnesium (Mg) | 2.4 | 4.8 | 6.7 | 9.5 |
| Sodium (Na) | 4.2 | 10. | 12. | 8,690. |
| Potassium (K) | 1.1 | 5.1 | 1.2 | 138. |
| Carbonate ($CO_3^{-2}$) | 0. | 0. | 0. | 3,010. |
| Bicarbonate ($HCO_3^-$) | 26. | 172. | 263. | 3,600. |
| Sulfate ($SO_4^{-2}$) | 12. | 8.0 | 5.4 | 10,500. |
| Chloride ($Cl^-$) | 5.0 | 5.0 | 20. | 668. |
| Fluoride ($F^-$) | 0.1 | 0.4 | 0.2 | |
| Nitrate ($NO_3^-$) | 3.2 | 20. | 1.3 | 5.8 |
| Dissolved solids | 64. | 250. | 310. | 25,000. |

[1]From *A Primer on Water Quality*. Geological Survey (1965).
[2]Stream in Connecticut
[3]Logan County, CO
[4]Drainage from the Everglades in Florida
[5]North-central North Dakota

drinking water. Industrial users usually require a higher quality water than this in order to prevent corrosion and boiler scale. On the other hand, certain irrigated crops can tolerate mineralized water as high as 2000 mg/L and more before crop growth is affected.

Dissolved solids are measured by filtering a known volume of sample and then evaporating the filtered water to dryness. The residue that remains is weighed and the results are recorded in milligrams per litre as filtrable residue (commonly called total dissolved solids, or TDS). This test is called the FILTRABLE RESIDUE TEST.

**Electrical conductivity.** A common way to determine the general dissolved solids or mineral content of water is to measure the ELECTRICAL CONDUCTIVITY (EC) of the water. Substances that dissolve in water are said to ionize—that is, to form electrically charged particles. These ionized particles allow water to conduct electricity—the more ionized particles (dissolved solids) in water, the greater the electrical conductivity.

The units of measure for EC are micromhos per centimeter at 25°C, abbreviated $\mu$mhos/cm at 25°C. As a general rule, every ten units of EC represents six to seven milligrams per litre of dissolved solids. Therefore, an EC of 1000 suggests a dissolved solids concentration of about 600-700 mg/L.[6] It should be noted that the relationship between EC and dissolved solids is different for every type of water.

EC is a quick test to conduct. It takes only a few minutes, whereas the filtrable residue test may take two hours to complete. The test is temperature-sensitive and is therefore always conducted at 25°C. Electrical conductivity is measured by using a conductivity meter like the one shown in Figure 17.

*Organic characteristics* is the second category of chemical characteristics. To date, more than 700 organic chemicals have been identified in drinking water supplies in the US. Organic chemicals have four general sources:

1. Plant and animal decomposition yield materials such as tannins, lignins, and fulvic and humic materials.

2. Wastewater discharges yield synthetic organics found in municipal wastes, chemical wastes, and industrial wastes.

3. Agricultural runoff yields synthetic organics such as pesticides and herbicides.

4. Water treatment operations yield complex organics, such as trihalomethanes (produced when water containing natural organic compounds is chlorinated).

There are two dangers posed by organics in drinking water. First, some organic materials are known to cause cancer, or are suspected of causing cancer, or have other adverse health effects. Second, some organic materials, perhaps

---

[6] *Basic Science Concepts and Applications,* Chemistry Section, Laboratory Test Calculations (Electrical Conductivity).

Courtesy of Balsbaugh Electrochemical Center,
Foxboro Analytical Division

**Figure 17. Typical EC Meter**

Courtesy of Johnson Controls, Inc.

**Figure 18. Biological Fouling**

harmless in their natural state, form complex organic compounds during chlorination, and the chlorinated organic complexes (trihalomethanes) are suspected as cancer-causing agents. In addition to these hazards, organic compounds can also cause tastes, odors, and color in water.

## Biological Characteristics

The term biological characteristics of water refers to the aquatic life, BACTERIA, and VIRUSES found in water. These characteristics may have a significant effect on the quality of water.

Algae, for example, cause taste and odor. Some types of algae clog sand filters; others produce nuisance slime growths on equipment, tanks, and reservoir walls (Figure 18). Several bacteria, known as iron and sulfur bacteria, cause corrosion in iron materials such as well screens and pipelines. Certain bacterial growths can clog screens, pumps, and valves. Others cause taste and odor problems. Microbiological life in water, such as bacteria, virus and PROTOZOA, can cause disease. A particular class of bacteria, known as the COLIFORM GROUP, is used as an indicator of the disease-causing potential of water. If laboratory results show that coliform bacteria are present, then it is probable that the water also contains organisms that can cause disease (PATHOGENS).

## Radiological Characteristics

RADIOACTIVE MATERIALS, or RADIONUCLIDES, in water supplies come from both natural and man-made sources. Naturally occurring radionuclides include radium-226 and radium-228. Radium is not usually present in significant amounts in surface waters. Radium can occur in ground waters, depending on the type of soil and rock material that the ground water contacts.

Man-made radioactivity can enter water supplies from a variety of sources that usually affect only surface water supplies. Fallout from nuclear weapons testing is perhaps the most significant source of man-made radioactivity. Other sources include medical, scientific, and industrial users of radionuclides, and EFFLUENTS (discharged waters) from nuclear power systems and from mining of radioactive materials.

## How Water Acquires Its Characteristics

Water acquires its characteristics from its surroundings—from the soils, rocks, minerals, and air that the water contacts, and from the animals and people that contact and influence the water. The following discussion focuses on the changing nature of surface- and ground-water characteristics as they are influenced by both the natural and man-made environments.

**Surface water quality.** Consider what can affect the quality of water from the time it originates high in the mountains until it reaches the raw-water storage reservoir.

Small mountain streams may carry clear water for most of the year. During storms, however, the water may carry moderate amounts of dirt, organic debris, and suspended material. Objectionable bacteria can always be present because the streams are exposed to organic wastes and disease-causing organisms from animals and humans.

As the streams and rivers move closer to inhabited areas, water quality can be further reduced. Agricultural runoff can add herbicides and pesticides. The water can pick up a wide variety of organics and inorganics from urban runoff. Inadequately treated domestic and industrial wastewaters can add toxic chemicals and disease-causing organisms.

The quality of the water reaching the raw-water storage reservoir may be relatively good, especially if the reservoir is located high in an uninhabited watershed; or it may be extremely poor, as it often is where the reservoir is located downstream of developed areas. In either case, the waters have been exposed to numerous opportunities for natural and man-made contamination. Surface waters should not be considered safe to use until they have undergone treatment.

The water in the storage reservoir is subject to additional changes in water quality. Although the water continues to be exposed to pollution and contamination, certain natural processes actually begin to improve water quality.

Cloudy or turbid water clears as the slow moving lake water allows suspended material to settle. Natural bacterial action OXIDIZES some of the non-settleable particles that might give rise to color, taste, and odor. Oxidation converts some of these particles into material that will settle.

By dilution, the sheer volume of the reservoir helps reduce water quality problems that are caused when flash floods or accidental spills deposit a relatively large amount of impurities into a small stream or river. Natural aeration (surface waters in contact with air and mixed by wind action), plus the oxygen produced by algae, maintains a dissolved oxygen level in the water. This

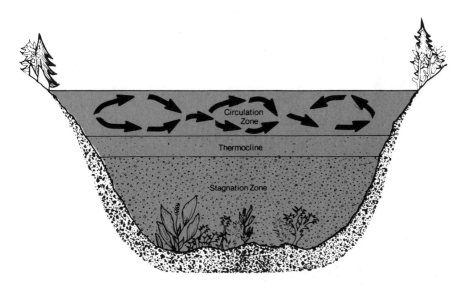

**Figure 19. Cross Section of Lake Showing Summer Stratification**

DO level supports a balanced and beneficial population of aquatic life (fish, aquatic plants, and bacteria).

Although certain beneficial bacteria can survive and even thrive in a reservoir, the reservoir is a hostile environment for many other organisms. Because of the relatively long time they remain in the reservoir (long detention time), many pathogens are destroyed.

Surface storage creates an excellent environment for algae to grow and thrive. Algae produce oxygen that improves the dissolved oxygen level of the lake, but certain species can also cause tastes and odor problems.

Temperature variations, winds, and the tendency of denser water to sink to the bottom act together to cause TURNOVER (the vertical circulation of the stored water) in large reservoirs. During the part of the year when the weather is warm, wind and temperature cause the lake to stratify into three layers: an upper layer of warm, well mixed, aerobic (oxygenated) water—the *circulating zone;* a middle, separating layer—the *thermocline;* and a cool, unmixed lower layer—the *stagnation zone* (Figure 19).

In late autumn (in all but the very warm areas of the US) the temperature in the circulation zone drops until it equals the temperature in the stagnation zone. At this time, wind action or further cooling of the circulation zone causes the surface waters to settle to the bottom. Water circulates throughout the reservoir, carrying oxygen to the depths and sediments and wastes to the surface. This is called the fall turnover.

Turnover can occur again in the spring. During the winter, water's unique property of being most dense at 39.2°F sets up another stratification (Figure

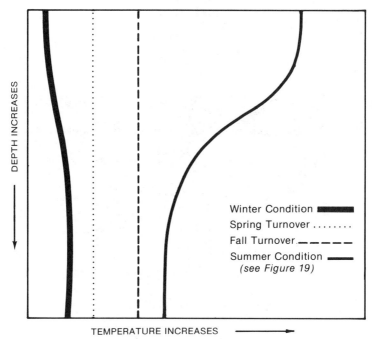

Winter Condition ▰▰▰▰
Spring Turnover ........
Fall Turnover _ _ _ _ _
Summer Condition ▬▬▬
*(see Figure 19)*

TEMPERATURE INCREASES    ⟶

**Figure 20.  How Temperature of Lake Water Changes With Depth**

Reprinted with permission of John Wiley & Sons, Inc., New York, N.Y., from *Elements of Water Supply
and Wastewater Disposal*, 2nd ed., by Gordon Fair, John Geyer, and Daniel Okun. Copyright ©1971.

20).[7] This time the lowest-density water at the surface (or directly below the ice)
is coldest [32° F (0° C)] and the heaviest bottom water is warmest [about 39.2° F
(4° C)]. In the spring, the lake surface warms and as it warms it approaches the
density of the lower water. Wind and density differences then cause the water to
mix throughout all levels. The process continues until the surface water reaches
39.2° F (4° C). The spring turnover (or spring overturn) is fairly rapid, but its
effects may be noticed for two weeks or longer. In warm climates, temperatures
throughout the reservoir may become the same, and the mixing action of the
wind alone can cause turnover. However, the effects of turnover are usually
more marked when it occurs in climates with greater seasonal temperature
variations.

Water treatment problems occur when bottom deposits and substances that
are normally present only in the stagnation layer are mixed throughout the
reservoir by turnover. In reservoirs with a thick thermocline (middle layer), light
cannot reach algae within the stagnation zone. With neither growing plants nor
regular contact with air (aeration) to supply oxygen, the stagnation zone
becomes a favorable environment for ANAEROBIC bacteria. The bacteria feed on
organic material and release foul-smelling and bad-tasting waste products,
which give water a characteristic septic taste and odor immediately after water

---

[7]*Basic Science Concepts and Applications,* Hydraulics Section, Density and Specific Gravity
(Density).

turnover. High concentrations of nutrients such as nitrogen and phosphorus are also brought to the surface during turnover, and these may trigger algae blooms (sudden, large populations). The algae can cause filter clogging in addition to more taste and odor problems.

Bottom water and bottom deposits are not the only source of nutrients. Runoff from agricultural areas and discharges from wastewater treatment plants can often have exceptionally high concentrations of nitrogen and phosphorus. Continued discharges of high nutrient wastes greatly add to *eutrophication* (the natural or man-made nutrient enrichment of a body of water).

Surface waters are also exposed to radioactive contamination from both natural and man-made sources. Naturally occurring radionuclides such as radium-226 and radium-228 are not as likely to be found in surface waters as in ground waters except in the vicinity of radioactive-materials mining operations. Man-made radionuclides can enter surface supplies as atmospheric fallout, from the effluents of nuclear power plants, or from the wastes of industries that use nuclear materials.

**Ground-water quality.** The quality of ground water is relatively uniform throughout an aquifer, and changes in quality occur slowly. Ground water is not exposed to the air, and is not as subject to direct pollution and contamination from runoff as surface water is.

Because of this protection, and because of the natural filtering which ground water receives as it passes through the aquifer, the biological (bacteriological) characteristics of ground water are generally very good (the natural filtering action of the aquifer removes most bacteria). In addition, the aquifer offers an unfavorable environment for disease-causing organisms to grow and multiply. Therefore, PATHOGENIC (disease-causing) organisms are rarely found in ground water. Most cases of ground-water contamination are a result of either inadequate well construction or poor waste disposal practices.

Once contaminated, an aquifer may be nearly impossible to reclaim because ground water does not benefit from natural purifying influences such as sunlight. And, because it moves so slowly, there is little quality improvement by dilution with other waters.

Ground water acquires its chemical characteristics in two ways:

• From surface waters that infiltrate and percolate into the aquifer, and

• From mineral deposits in the aquifer, some of which are dissolved by ground water.

In order for infiltration/percolation to occur, there has to be a water source, and there must be soil or rock conditions that will permit the downward movement of water. When both conditions exist, water from a surface source can reach the aquifer. The water source may be a clean, good quality stream; this kind of ground-water recharge could improve the quality of the ground water. However, the percolating waters may be polluted; for instance, waters that percolate through fertilized agricultural areas could have high nitrate concentrations, and CONTAMINATION from herbicides and pesticides.

As ground water passes through the aquifer material, it dissolves some of the minerals and substances that it contacts. Because it is in contact with the aquifer materials for long periods, ground water tends to have a high mineral content. Depending on the minerals in the aquifer, the water may have high concentrations of iron and manganese, which can cause taste, odor, and color. The water may also be very hard due to dissolved calcium and magnesium. Often, ground waters in areas having both iron and sulfur deposits will have the characteristic rotten egg odor of hydrogen sulfide. Some of the common chemical characteristics of ground water are

| Iron | Nitrate |
|---|---|
| Manganese | Sodium |
| Fluoride | Magnesium |
| Chloride | Calcium |
| Sulfate | Hydrogen sulfide |

The radiological characteristics of ground water can come from surface waters that infiltrate and percolate into the aquifer. However, it is more likely that radionuclides will be introduced naturally as ground water moves through and dissolves radioactive mineral deposits.

## 1-5. Public Health Significance of Water Quality

An increasingly large part of the operator's responsibility during recent years has been to ensure that the water system complies with state and federal water-quality standards. You should understand what these standards are and why they were chosen.

### The Interim Primary Regulations (IPRs)

The Interim Primary Regulations of the Safe Drinking Water Act (SDWA) were developed to protect the public's health. The regulations, which went into effect June 24, 1977, establish MAXIMUM CONTAMINANT LEVELS (MCLs). These levels, given in tables 3, 4, 6, and 7, place limits on certain impurities that are known to be found in water and that are known to cause illness or disease. Appendix A contains a table summarizing the Interim Primary Regulations.

### Physical Characteristics

Water temperatures have no direct effect on the public's health. However, a sudden change in the temperature of a water supply warrants investigation. Perhaps the change merely indicates that a lake has undergone its seasonal turnover, or perhaps it indicates that there has been a massive discharge of wastewater immediately upstream.

Color, taste, and odor, though primarily aesthetic concerns, can be indicators of pollution from sources such as domestic or industrial wastes.

Turbidity is a public health concern because it interferes with the disinfection process. Generally, chlorine is used for disinfection. Its effectiveness depends

**Table 3.  Maximum Contaminant Levels: Turbidity**

**A.** The *monthly average turbidity* may not exceed 1 TU. (A 5-TU monthly average may apply at state option, provided it does not interfere with disinfection, maintenance of an effective chlorine residual, or bacteriological testing.)

**and**

**B.** The *average turbidity for two consecutive days* may not exceed 5 TU.

upon *contact* between the disinfectant and the organisms to be destroyed. Turbidity particles can provide hiding places for organisms where the disinfectant cannot reach. The "hidden" organisms can then pass through the treatment process, reach the consumer, and possibly cause disease. Because of the public health significance, the Interim Primary Regulations include maximum turbidity levels, summarized in Table 3.

## Chemical Characteristics

The inorganic and organic chemicals singled out by the Interim Primary Regulations as being health hazards are shown in Table 4. This summary gives the health effects and the MCLs, and tells whether each undesirable health effect results from long-term or short-term exposure to the chemical. Further research is under way to identify and set requirements for other harmful organic chemicals. (MCLs and required treatment processes for *trihalomethanes* were pending final approval when this book was published.)

In addition to the data in Table 4, you will need some background information to understand how the regulations deal with four chemical characteristics of water—nitrate, fluoride, sodium, and sulfate.

*Nitrate* ($NO_3$) has caused methemoglobinemia (infant cyanosis or "blue baby disease") in infants who have consumed water or formulas prepared with water having high nitrate concentrations. The Interim Primary Regulations state that a domestic water supply should not contain nitrate concentrations in excess of 10 mg/L as nitrogen (about 45 mg/L expressed as nitrate).[8] The nitrate concentration should be determined, and, if the concentration is excessive, water should not be used for feeding infants, and advice should be obtained from health authorities about the suitability of using the water for drinking purposes.

In some polluted wells, nitrite ($NO_2$) will also be present in concentrations greater than 1 mg/L; this is even more hazardous to infants than is nitrate. When such a nitrite concentration is suspected, the water should *not* be used for infant feeding.

It is an established fact that *fluoride* in water, to the extent of about 1 mg/L, will help prevent tooth decay in children. However, excessive amounts result in a brown discoloration (when the fluoride exceeds 2.5 mg/L) and pitting of the teeth (when the fluoride exceeds 6 mg/L) called mottling, or dental fluorosis. The condition is particularly noticeable in long-time residents of communities

[8] *Basic Science Concepts and Applications,* Chemistry Section, Solutions (Normality).

**Table 4.   Chemicals and Their Health Effect**

| Contaminant | Health Effect* | Maximum Contaminant Level mg/L |
|---|---|---|
| Arsenic | Small sores on hands and feet, possibly developing into cancers | 0.05 |
| Barium | Increased blood pressure and nerve block | 1. |
| Cadmium | Concentrates in liver, kidneys, pancreas, and thyroid; hypertension is a suspected health effect | 0.010 |
| Chromium | Skin sensitization, kidney damage | 0.05 |
| Lead | Constipation, loss of appetite, anemia; tenderness, pain and gradual paralysis in the muscles, especially the arms | 0.05 |
| Mercury | Inflammation of the mouth and gums; swelling of the salivary glands; loosening of teeth | 0.002 |
| Selenium | Red staining of fingers, teeth, and hair; general weakness; depression; irritation of the nose and throat | 0.01 |
| Silver | Permanent gray discoloration of skin, eyes, and mucous membranes | 0.05 |
| Fluoride | Stained spots on teeth (mottling)—the amount of discoloration depends on the amount of fluoride ingested | Based on annual average of maximum daily air temperature:<br><br>°F      MCL<br>53.7 & below: 2.4<br>53.8 to 58.3:  2.2<br>58.4 to 63.8:  2.0<br>63.9 to 70.6:  1.8<br>70.7 to 79.2:  1.6<br>79.3 to 90.5:  1.4 |
| Nitrate (as N) | Temporary blood disorder in infants—can be fatal | 10. |
| Pesticides<br>Endrine<br>Lindane<br>Methoxychlor<br>Toxaphene | Cause symptoms of poisoning which differ in intensity. The severity is related to the concentration of these chemicals in the nervous system, primarily the brain. Mild exposure causes headaches, dizziness, numbness and weakness of the extremities. Severe exposure leads to spasms involving entire muscle groups, leading in some cases to convulsions. Suspected of being carcinogenic. | 0.0002<br>0.004<br>0.1<br>0.005 |
| Herbicides<br>2,4-D and<br>2,4,5-TP | Liver damage and gastrointestinal irritation | 0.1<br>0.01 |

*The health effects of nitrate result from short-term exposure. All other health effects result from exposure over a long period of time.

Table 5.    Waterborne Diseases

| Waterborne Disease | Causative Organism | Source of Organism in Water | Health Effect |
|---|---|---|---|
| Gastroenteritis | Various pathogens | Animal or human feces | Acute diarrhea and vomiting |
| Typhoid | Salmonella typhosa (bacteria) | Human feces | Inflamed intestine, enlarged spleen, high temperatures—fatal |
| Bacillary dysentery | Shigella (bacteria) | Human feces | Diarrhea, rarely fatal |
| Cholera | Vibrio comma (bacteria) | Human feces | Vomiting, severe diarrhea, rapid dehydration, mineral loss— high mortality |
| Infectious hepatitis | Virus | Human feces | Yellow skin, enlarged liver, abdominal pain —low mortality—lasts up to 4 months |
| Amebic dysentery | Entamoeba histolytica (protozoa) | Human feces | Mild diarrhea, chronic dysentery |
| Giardiasis | Giardia lamblia (protozoa) | Wild and domestic animal feces | Diarrhea, cramps, nausea and general weakness—lasts 1 week to 30 weeks— not fatal |

served by waters that are high in fluoride.

The beneficial or adverse effects of fluoride occur as a direct result of how much fluoride is taken into the body. This depends on two things: (1) the fluoride concentration in water; and (2) the amount of water consumed. The amount of water we drink is directly related to the outside air temperature since we tend to drink more when it is hot. Because of this, the Interim Primary Regulations contain MCLs for fluoride based on the annual average of maximum daily air temperatures, as shown in Table 4.[9]

For healthy persons, the *sodium* content of water is unimportant, because the intake from salt and foods is so much greater than the intake from water. For this reason, the Interim Primary Regulations do not establish an MCL for sodium. However, for persons placed on low-sodium diets because of heart, kidney, or circulatory ailments, or because of complications of pregnancy, the sodium content of water must be considered. Also, when home water softeners utilizing the ion exchange method are used, the amount of sodium in the

---

[9]*Basic Science Concepts and Applications,* Chemistry Section, Laboratory Test Calculations (Fluoride).

**Table 6.    Coliform Bacteria Maximum Contaminant Levels**

*Membrane Filter Method*

**A.**

*For all systems,* regardless of the number of samples taken per month, the coliform level must not exceed 1 coliform colony per 100 ml for the average of all monthly samples.

**and B.**

| | |
|---|---|
| *For all systems required to take fewer than 20 samples per month,* the coliform level must not exceed 4 coliform colonies per 100 ml in more than one of the monthly samples taken. | *For systems required to take 20 or more samples per month,* the coliform level must not exceed 4 coliform colonies per 100 ml in more than 5 percent of the monthly samples taken. |

*Multiple Tube Fermentation Methods (10 ml portions)*

**A.**

*For all systems,* regardless of the number of samples taken per month, coliform must not be present in more than 10 percent of the portions per month.

**and B.**

| | |
|---|---|
| *For systems required to take fewer than 20 samples per month,* not more than 1 monthly sample can have 3 or more portions positive. | *For systems required to take 20 or more samples per month,* not more than 5 percent of the monthly samples can have 3 or more portions positive. |

consumer's treated water is increased; for this reason, water that has been softened should be analyzed for sodium if a precise record of individual sodium intake is recommended. The usual low-sodium diets allow for 20 mg/L sodium in the drinking water. If this limit is exceeded, persons on sodium-controlled diets should seek a physician's advice on diet and sodium intake.

Water containing high concentrations of *sulfate,* caused by the dissolving of natural deposits of magnesium sulfate (Epsom salts) or sodium sulfate (Glauber's salt) may be undesirable because of their laxative effects. No MCLs are established for sulfates since the laxative effect seems to be experienced only by newcomers and occasional users of water that is high in sulfates. Evidently the body becomes accustomed to using high sulfate waters in a relatively short time. It is recommended, however, that the sulfate content of water not exceed 250 mg/L.

## Biological Characteristics

Disease-causing organisms (pathogens) can be transmitted in water. WATERBORNE DISEASES can be caused by bacteria, viruses, and protozoa. Table 5 summarizes the more common waterborne diseases and identifies both the organism that causes the disease and the disease symptoms.

Disease-causing organisms are found in the intestinal tract of humans and other warm-blooded animals. Since coliform bacteria are *always* present in the normal intestinal tract, coliform bacteria have been the bacteriological tool used to indicate fecal contamination of water. When coliform bacteria are present in a

**Table 7.  Radiological Contaminants and Their Health Effects**

| Contaminant | Health Effect* | Screening Procedure | Maximum Contaminant Level |
|---|---|---|---|
| Natural<br><br>Gross Alpha<br>Combined radium-226<br>and radium-228 | Bone cancer | (1) Test for gross alpha activity. If gross alpha *exceeds* 5 pCi/L, then:<br>(2) Test for radium-226. If radium-226 *exceeds* 3 pCi/L, then:<br>(3) Test for radium-228 | Gross Alpha: 15 pCi/L†<br><br>Radium-226<br>    +    :   5 pCi/L<br>Radium-228 |
| Man-made**<br><br>Gross Beta<br>Tritium<br>Strontium-90 | Cancer | (1) Test for gross beta, tritium, and strontium-90. If the following screening levels are not exceeded, no further testing is required:<br>    Gross Beta: 50 pCi/L<br>    Tritium: 20,000 pCi/L††<br>    Strontium-90: 8 pCi/L††<br>    If the screening level for gross beta is exceeded, then:<br>(2) Perform a complete analysis to identify the major radioactive constituents and the appropriate organ and total body doses. | 4 millirem/year† |

*Health effects result from long-term exposure.
†See glossary for definition.
**These MCLs apply only to surface water systems serving 100,000 people or more.
††If tritium or strontium-90 exceeds the screening levels, the MCL is exceeded. If *both* radionuclides are present, the sum of their annual dose equivalents to bone marrow shall not exceed 4 millirem/year.

water, it is assumed that disease-causing organisms may also be present.

One of two methods is used for detecting coliform bacteria: (1) the membrane filter method or (2) the multiple-tube fermentation method. The MCLs established by the Interim Primary Regulations for coliform bacteria are given in Table 6 for each testing method.

## Radiological Characteristics

The effects of human exposure to radiation or radioactive materials are harmful, and exposure should be avoided. Man has always been exposed to natural radiation from water, food, and air. For example, radioactive material in rivers may collect in microscopic life forms that serve as food for fish and other forms of life that can be consumed by man.

With the development and use of radioactive materials for energy, defense, and industry, the likelihood of exposure has increased dramatically. Exposure

**Table 8.    Secondary MCLs**

| Contaminant | Level |
|---|---|
| Chloride | 250 mg/L |
| Color | 15 color units |
| Copper | 1 mg/L |
| Corrosivity | Non-corrosive |
| Foaming agents | 0.5 mg/L |
| Hydrogen sulfide | 0.05 mg/L |
| Iron | 0.3 mg/L |
| Manganese | 0.05 mg/L |
| Odor | 3 threshold odor number |
| pH | 6.5–8.5 |
| Sulfate | 250 mg/L |
| Total dissolved solids | 500 mg/L |
| Zinc | 5 mg/L |

to either natural or man-made radiation can cause genetic defects and cancer. Excessive exposure can be fatal. The exposure to radiation from water alone is not likely to be fatal, but it does add to the total exposure.

The Interim Primary Regulations establish safe levels for natural and man-made radionuclides. The maximum contaminant levels are given in Table 7.

## The Secondary Regulations

The Proposed Secondary Regulations of the Safe Drinking Water Act set maximum levels for contaminants in public drinking water (a) which may adversely affect the odor or appearance of the water, causing a large number of people to discontinue its use, or (b) which may otherwise adversely affect the public welfare.

The levels set by the Proposed Secondary Regulations should give you an idea of what standards are recommended for quality drinking water in addition to those required by the Interim Primary Regulations. The maximum contaminant levels of the Proposed Secondary Regulations are shown in Table 8.

## Selected Supplementary Readings

### The Water Cycle—The Fresh Water Delivery System

*A Primer on Water.* Geological Survey, US Dept. of Interior, (1960). pp. 29-31.
*Manual of Instruction for Water Treatment Plant Operators.* New York State Dept. of Health (no date). pp. 3-5.
*Water Distribution Operator Training Handbook.* American Water Works Association (1976). p. I-1.
*Manual of Individual Water Supply Systems.* Water Supply Div., Environmental Protection Agency, reprint 1975. pp. 1-3.
*Ground Water and Wells.* Johnson Division, UOP Inc., revised edition (1975). pp. 15-16.
Linsley, Ray K.; Kohler, Max A.; & Paulhus, Joseph L.H. *Hydrology for Engineers.* McGraw-Hill Book Company (1958). pp. 1-5.
Peixoto, J.P.; & Kettani, M. Ali. The Control of the Water Cycle, Scientific American: 228:46-61 (Apr. 1973).

### Surface Water—Water on the Land

*A Primer on Water.* Geological Survey, US Dept. of Interior (1966). pp. 5, 19-29.
Murray, C.R.; & Reeves, E. Bodette. *Estimated Use of Water in the United States in 1975.* Geological Survey Circular 765 (1977). pp. 20 & 33.
Linsley, R.K.; & Franzini, J.B. *Water-Resources Engineering.* McGraw-Hill Book Co. (1964). pp. 135-137, 139, 140.
*Manual of Instruction for Water Treatment Plant Operators.* New York State Dept. of Health. pp. 5-10.
*Manual of Individual Water Supply Systems.* Water Supply Div., Environmental Protection Agency. pp. 1, 4, 5, 61-71.
*Manual of Water Utility Operations.* Texas Water Utility Assn. (1975). pp. 40-41, 53-54.

### Ground Water—The Hidden Resource

*A Primer on Ground Water.* Geological Survey, US Dept. of Interior (1963). pp. 5-10.
*Ground Water and Wells.* Johnson Division, UOP Inc., revised edition (1975). pp. 5-11, 13, 15-22.
*Manual of Instruction for Water Treatment Plant Operators.* New York State Dept. of Health, pp. 10-14.
*Manual of Individual Water Supply Systems.* Water Supply Div., Environmental Protection Agency. pp. 3-5.
*Manual of Water Utility Operations.* Texas Water Utilities Assn. (1975). pp. 19-21.

## Characteristics of Water

*Manual of Instruction for Water Treatment Plant Operators.* New York State Dept. of Health. pp. 45-47, 257-259, 262-263, 264-265, 267-268, 277-278, 289, 292, 296.

*Manual of Individual Water Supply Systems.* Water Supply Div., Environmental Protection Agency. pp. 5-14.

*Manual of Water Utility Operations,* Texas Water Utilities Assn. (1975). pp. 12-13, 56-58.

Sawyer, Clair N. & McCarty, Perry L. *Chemistry for Sanitary Engineers.* McGraw-Hill Book Company (1967). pp. 290-292, 299-300, 314-316, 347-351, 383-386, 435-436.

## Surface Water Quality

*Manual of Water Utility Operations,* Texas Water Utilities Assn. (1975). pp. 40-43, 48-52.

Overman, Michael. *Water, Solutions to a Problem of Supply and Demand.* Doubleday and Co., Inc. (1969). pp. 80-81.

Davis, Calvin V., ed. in chief; Sorenson, Kenneth E., coeditor. *Handbook of Applied Hydraulics.* McGraw-Hill Book Company (1942). pp. 687-692.

Hirsch, A.A. *Manual for Water Plant Operators.* Chemical Publishing Co., Inc. (1945). pp. 42-46.

## Ground Water Quality

Gibson, Ulric P. & Singer, Rexford D. *Water Well Manual.* Premier Press (1971). pp. 22-27.

*Manual of Individual Water Supply Systems.* Water Supply Div., EPA reprint (1975). pp. 5-14.

*Manual of Water Utility Operations.* Texas Water Utilities Assn. (1975). pp. 34-39.

## Public Health Significance of Water Quality

*Foodborne and Waterborne Disease Outbreaks.* Center for Disease Control, US Dept. of Health Education and Welfare. Annual Summary (1973, 1974, and 1975).

Craun, Gunther F.; McCabe, Leland J.; & Hughes, James M. *Waterborne Disease Outbreaks in the U.S.—1971-1974. Jour. AWWA,* 68:8:420 (Aug. 1976).

*Community Water Supply Study, Analysis of National Survey Findings.* US Dept. of Health, Education and Welfare; US Public Health Service; Environmental Health Service; and Bureau of Water Hygiene (Jul. 1970).

# Glossary Terms Introduced in Module 1

(Terms are defined in the Glossary at the back of the book.)

Acidic water
Aeration
Aesthetic
Algae
Alkaline water
Anaerobic
Appropriative rights
Aquatic life
Aquifer
Artesian aquifer
Artesian well
Atom
Bacteria
Basic Water
Capillary action
Coliform-group bacteria
Color
Color unit (CU)
Condensation
Contamination
Corrode (corrosion)
Dissolved oxygen (DO)
Dissolved solids
Divide
Drainage basin
Drainage divide
Effluents
Electrical conductivity (EC)
Ephemeral stream
Evaporation
Filtrable residue test
Ground water
Ground-water aquifer
Hardness
Impermeable layer
Infiltration
Inorganic material
Maximum contaminant level
  (MCL)
Milligrams per litre (mg/L)

Millirem
Molecule
Organic material
Oxidize
Palatable (palatability)
Parts per million (ppm)
Pathogens (pathogenic)
pCi/L
Percolation
Perennial stream
pH
Photosynthesis
Piezometric surface
Pollution
Precipitation
Protozoa
Radioactive materials
Radionuclide
Rainfall intensity
Recharge
Riparian rights
Saturated
Surface runoff
Surface water
Temperature
Threshold odor
Transpiration
Turbidity
Turbidity unit (TU)
Turnover
Virus
Void space
Water body
Waterborne disease
Water course
Watershed
Watershed divide
Water table
Water table aquifer
Water table well

# Review Questions

(Answers to Review Questions are given at the back of the book.)

1. Define water cycle.

2. List the seven major elements that make up the water cycle.

3. What is the process called that changes water vapor to tiny droplets of water or tiny ice crystals?

4. Explain the difference between infiltration and percolation.

5. Which of the two processes, infiltration or percolation, makes water available to the roots of vegetation?

6. List seven factors that influence the amount of surface runoff.

7. What happens when certain soils, like clay, become wet? How does this affect infiltration?

8. The boundary or ridge that separates one drainage basin from another is called a _____.

9. What are the names given to *year-round* and *intermittently flowing* streams?

10. What are the probable sources of surface water for year-round streams?

11. Name the two basic water rights doctrines.

12. The top of the zone of saturation is called _____.

13. Define the term *aquifer*.

14. List the two general types of aquifers.

15. List the four general categories of water quality characteristics.

16. What are the preferred units of measure for concentration?

17. Why is temperature an important consideration from a water supply and treatment standpoint?

18. Make the following temperature conversions between Fahrenheit and Celsius:
    (a) $100°C$ = _____ $°F$
    (b) $-10°C$ = _____ $°F$
    (c)  $32°F$ = _____ $°C$
    (d)  $68°F$ = _____ $°C$
    (e) $39.2°F$ = _____ $°C$

    Describe the condition of water at $100°C$, $32°F$, and $39.2°F$.

19. _____is a measurement of the cloudiness of water.

20. Which of the following are considered chemical characteristics of water?
    (a) pH
    (b) hardness
    (c) color
    (d) turbidity
    (e) dissolved oxygen

21. List two problems associated with the use of hard water.

22. What are the equivalent weights of the following:
    (a) calcium (Ca)
    (b) magnesium (Mg)
    (c) calcium carbonate ($CaCO_3$)

23. Laboratory test results reported that a water sample had a hardness of 120 mg/L as Ca. What is the hardness expressed as $CaCO_3$?

24. What are the units of measure for electrical conductivity?

25. The laboratory determined that a particular water sample had an electrical conductivity of 700 $\mu$mhos/cm. What total dissolved solids concentration does this suggest?

26. What are some of the ways in which a reservoir can help improve water quality?

27. Is there a danger or risk involved in drinking water from a crystal clear mountain stream or lake? Explain.

28. Describe "turnover."

29. At what temperature does water reach its maximum density?

30. Explain why the following statement is true or false: *Ground-water quality varies widely within an aquifer and can change very quickly.*

31. List at least five chemical substances that are generally present in ground water.

32. Which of the physical characteristics interferes with disinfection by chlorination and how does this occur?

33. Name at least five chemicals found in water which can cause harmful health effects in man.

34. A laboratory reports a particular nitrate concentration to be 10 mg/L *as N*. What is this nitrate concentration expressed in mg/L *as* $NO_3$?

35. Why is the maximum allowable fluoride concentration expressed in terms of air temperature?

36. The following table summarizes the average daily maximum air temperatures for Fallbrook, California. (a) Calculate the annual average of these air temperatures. (b) Using Table 4 in your book, determine which fluoride maximum contaminant level would be applicable for Fallbrook.

| Month | Average Daily Maximum °F | Month | Average Daily Maximum °F |
|---|---|---|---|
| January | 68.4 | July | 85.3 |
| February | 69.2 | August | 87.3 |
| March | 67.3 | September | 84.3 |
| April | 70.4 | October | 82.0 |
| May | 72.6 | November | 72.4 |
| June | 76.8 | December | 70.0 |

37. List at least five microbiological diseases which can be transmitted by water.

## Study Problems and Exercises

1. (a) On a map of your area, locate and outline the watershed or watersheds that serve your water supply system.

   (b) Label all water courses and water bodies which are tributary to the raw water supply. Also, label all major physical features and land uses (such as agricultural, industrial, commercial, and residential) within the basin.

   (c) Locate and label all water wells used by the water supply system.

   (d) Plot and label all sources of pollution. List the watershed protection measures currently in force.

2. Is your water supply system entitled to *all* the water in the watershed(s) or is the water shared? Explain. Show the amount of water to which each user is entitled, and list the users in the order of their priority to the water.

3. Find out what type of water right(s) your water supply system has. Is (are) the right(s) based on the riparian or the appropriative doctrine?

4. (a) From one month's laboratory records at a water supply system, calculate average values for the following characteristics:

   Temperature                          pH (range, not average)

   Turbidity                            Hardness

   Color                                Total Dissolved Solids

   Odor

   (b) On the same tabulation, compare these values with the recommended or maximum allowable levels established under the Interim Primary and Proposed Secondary Regulations of the Safe Drinking Water Act. (Turbidity, color, pH, and dissolved solids only.)

5. Prepare a report on a waterborne disease outbreak in your city or region. Your local or state health agency may be able to assist in identifying recent outbreaks and sources of information. The report should identify

   • Location of outbreak

   • Type of disease

   • Number of people affected

   • Disease-causing organism

   • Suspected or confirmed cause of contamination

   • Action taken to prevent recurrence

6. (a) Based on the most current data on the quality of ground water *delivered* to the consumer, list the average concentrations of the following constituents:

   Arsenic              Lead              Silver

   Barium               Mercury           Flouride

   Cadmium              Selenium          Nitrate

   (b) Compare the individual concentrations (not the averages) against the limits established by the Interim Primary Regulations as listed in Table 4.

   (c) Explain the corrective actions that are being taken to bring any nonconforming characteristics into compliance—or steps being taken to expand the monitoring program to include sampling for constituents not yet being monitored.

# Water Use

The amount of water people use varies widely from place to place, depending on whether it is used for public water supplies, for agriculture, for rural uses (privately owned rural water supplies), or by self-supplied industry. Water use varies widely within these four catagories, depending on factors such as climate, season, quality, pressure, and cost.

As an operator, you are concerned primarily with the *amount* of water needed for public water supplies since this determines the amount of water that the treatment plant and distribution system must deliver. You are also concerned with the *rate* at which water must be delivered. The rate is determined by factors such as climate, type of water use, and the habits of the community.

By understanding the factors that control water use, you will be better able to plan for and meet the water use needs of your community. After completing this module you should be able to

- List the four principal water uses

- Describe the four public water use categories

- Explain the ten factors that affect the amount and rate of public water use and identify which factors the operator can control

- Explain the measurements of water use commonly made in treatment plants

## 2-1. How Water Is Used

In the US during 1975, an average of 420 billion gallons of water was withdrawn from water sources each day (420 bgd) for four principal uses:

- Public water supplies..................................... 29 bgd

- Irrigation................................................140 bgd

- Rural uses............................................... 4.9 bgd

- Self-supplied industry ...................................240 bgd*

*Figures do not add to total due to rounding.

Table 9.   Water Used for Public Supplies, by States, 1975*

| State | Gallons per Capita per day | State | Gallons per Capita per day |
|---|---|---|---|
| Alabama | 210 | New Mexico | 236 |
| Alaska | 442 | New York | 154 |
| Arizona | 211 | North Carolina | 169 |
| Arkansas | 139 | North Dakota | 130 |
| California | 185 | Ohio | 167 |
| Colorado | 200 | Oklahoma | 130 |
| Connecticut | 134 | Oregon | 190 |
| Delaware | 171 | Pennsylvania | 178 |
| Florida | 168 | Rhode Island | 128 |
| Georgia | 158 | South Carolina | 242 |
| Hawaii | 228 | South Dakota | 126 |
| Idaho | 236 | Tennessee | 130 |
| Illinois | 199 | Texas | 176 |
| Indiana | 146 | Utah | 331 |
| Iowa | 146 | Vermont | 150 |
| Kansas | 170 | Virginia | 119 |
| Kentucky | 101 | Washington | 256 |
| Louisiana | 152 | West Virginia | 154 |
| Maine | 143 | Wisconsin | 156 |
| Maryland | 147 | Wyoming | 191 |
| Massachusetts | 145 | District of Columbia | 215 |
| Michigan | 168 | Puerto Rico — Virgin Islands | 125 |
| Minnesota | 135 | | |
| Mississippi | 120 | | |
| Missouri | 158 | | |
| Montana | 267 | | |
| Nebraska | 248 | | |
| Nevada | 321 | | |
| New Hampshire | 115 | | |
| New Jersey | 145 | United States | 168 |

*Estimated Use of Water in the U.S. in 1975. Geological Survey Circular 765 (1977).

In the operation of water treatment plants you are primarily concerned with only one of the four principal uses—public water supplies. The 29 bgd withdrawn for public water supply in 1975 represents an average of 168 gallons per day for each individual served (168 GALLONS PER CAPITA PER DAY, or 168 gpcd). The average rate of water use per person varies widely from one location to another, depending on factors such as the climate and character of the community. Table 9 gives a state-by-state comparison of average water use per capita (that is, by each person). As you can see, public water supply use varied from a low of 101 gpcd in Kentucky to a high of 442 gpcd in Alaska.[10]

---

[10]*Basic Science Concepts and Applications,* Mathematics Section, Per Capita Water Use.

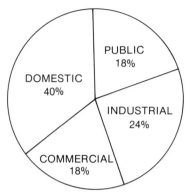

**Figure 21.  US National Average Public Water Supply Use**

## Public Water Supply Use Categories

All public water supply uses can be grouped into four categories, as shown in Figure 21. By applying the percentages in the figure to 168 gpcd you can determine what the average per capita use in each of the four categories was for 1975:

- Domestic uses, 68 gpcd

- Commercial uses, 30 gpcd

- Industrial uses, 40 gpcd

- Public uses, 30 gpcd

Of the four categories, domestic uses take the most water.

**Typical domestic uses.**    Activities using water in and around the home are DOMESTIC WATER USES. Some typical ones, with their approximate amounts are

Laundering . . . . . . . . . . . . . . . . . . . . . . . . . . . . . 20 to 45 gallons per load

Shower . . . . . . . . . . . . . . . . . . . . . . . . . . . . . . . . . . . 20 to 30 gallons

Tub bath . . . . . . . . . . . . . . . . . . . . . . . . . . . . . . . . . 30 to 40 gallons

Dishwashing . . . . . . . . . . . . . . . . . . . . . . . . . . . 15 to 30 gallons per load

Toilet . . . . . . . . . . . . . . . . . . . . . . . . . . . . . . . . 3.5 to 7 gallons per flush

Garbage disposal* . . . . . . . . . . . . . . . . . . . . . . . . . . . 5 gallons per minute

Lawn watering** . . . . . . . . . . . . . 7 to 43 gallons per week for each 100 sq ft

Car washing* . . . . . . . . . . . . . . . . . . . . . . . . . . . . . 5 gallons per minute

Drinking water . . . . . . . . . . . . . . . . . . . . . 1 to 2 quarts per day per person

---

*Assuming that faucet normally draws 5 gallons per minute fully opened.
**Use the lower value in humid areas, the high value in arid areas, and a mid-range value in semi-arid regions. Excessive or wasteful practices will greatly exceed these numbers.

In arid and semiarid regions of the country, lawn and garden irrigation takes more water than any other domestic use.

Notice that, although public water systems treat all domestic water to raise it to drinking-water quality, only a fraction (less than 1 percent) is actually used for drinking. However, users expect water of drinking-water quality for other household purposes, such as laundering, showering, bathing, and dishwashing.[11]

**Typical commercial uses.**    Today's public water system serves a variety of commercial water users, such as

| | |
|---|---|
| Motels | Car washes |
| Office buildings | Service stations |
| Shopping centers | Airports |
| Laundries | Swimming pools |

In addition to domestic uses found in motels, airports and offices, COMMERCIAL WATER USES include water for heating, cooling, humidifying, washing, ice making, vegetable and produce watering, ornamental fountains, and landscape irrigation.

**Typical industrial uses.**    The amounts of water used by different industries vary widely, depending on the type of industry, and on water availability, water cost, wastewater disposal problems, types of processes and equipment, and water conservation and reuse practices.

Some industries find it more economical to purchase water from the public system; others develop their own supplies. Historically, the major industrial water users, such as the steel, petroleum products, pulp and paper, coke, and power industries, have provided their own water supplies; smaller industries and industries with low water usage have purchased water from public water systems.

Here are some major INDUSTRIAL WATER USES:

| | |
|---|---|
| Heating | Pollution control |
| Cooling | Dust control |
| Conveying materials | Water sold as part |
| Washing | of the product |

Water used by industry is usually measured in terms of the product produced.[12] Table 10 gives water requirements for some typical industries.

**Public water uses.**    Municipalities and other public bodies own facilities and provide services that require water. Typical publicly owned facilities are

| | |
|---|---|
| Public parks and golf courses | Fairgrounds |
| Campgrounds | Municipal buildings |
| Recreational areas | Municipal swimming pools |
| Hospitals | and beaches |

---

[11] *Basic Science Concepts and Applications,* Mathematics Section, Domestic Water Use Based on Household Fixture Rates.

[12] *Basic Science Concepts and Applications,* Mathematics Section, Water Use per Unit of Industrial Product.

Table 10.    Water Requirements in Selected Industries*

| Process | Water Required (in gallons per unit of product noted) |
|---|---|
| *Canneries:* | |
| Green beans, per ton | 20,000 |
| Peaches and pears, per ton | 5,300 |
| Other fruits and vegetables, per ton | 2,000–10,000 |
| *Chemical industries:* | |
| Ammonia, per ton | 37,500 |
| Carbon dioxide, per ton | 24,500 |
| Gasoline, per 1,000 gal | 7,000–34,000 |
| Lactose, per ton | 235,000 |
| Sulfur, per ton | 3,000 |
| *Food and beverage industries:* | |
| Beer, per 1,000 gal | 15,000 |
| Bread, per ton | 600–1,200 |
| Meat packing, per ton live weight | 5,000 |
| Milk products, per ton | 4,000–5,000 |
| Whiskey, per 1,000 gal | 80,000 |
| *Pulp and paper:* | |
| Pulp, per ton | 82,000–230,000 |
| Paper, per ton | 47,000 |
| *Textiles:* | |
| Bleaching, per ton cotton | 72,000–96,000 |
| Dyeing, per ton cotton | 9,500–19,000 |
| *Mineral products:* | |
| Aluminum (electrolytic smelting), per ton | 56,000 (max) |
| *Petroleum,* per barrel of crude oil | 800–3,000 |
| *Steel,* per ton | 1,500–50,000 |

*Sources:
Metcalf and Eddy, *Wastewater Engineering*, McGraw-Hill Book Company, 1972, p. 32.
Fair, Geyer and Okun, *Elements of Water Supply and Wastewater Disposal*, John Wiley Sons, 2nd Ed., 1971, p. 31.

These facilities require water for drinking, irrigating, washing, cleaning, recreation, and decoration (lakes, ponds, reflecting pools, and ornamental fountains). Such uses are PUBLIC WATER USES.

In addition, municipalities provide vital services, such as street cleaning, water line cleaning, and sewer line cleaning. An important public service is fire fighting. Although the total annual volume of water required for fire fighting is small, the rate of flow required during a fire can be very large. For a large fire it can be the largest single demand a water system will ever need to meet. For example, a community that uses an average of 1 mgd (1 million gallons per day) might reasonably expect a fire flow demand of 2500 gpm (2500 gallons per minute) for up to 10 hours; this flow would use 3.6 mgd if it continued for a full day—three and one-half times the average daily flow.

## 2-2.  Variations in Water Use

The amount of water used varies with the time of day, day of the week, and season of the year. The amount of water used and the variations in that use are partly controlled by known factors. Understanding these factors will help you to understand your community's water needs, and help you to anticipate when demands can be expected, why they may be excessively high, and how they might be reduced.

### Factors Affecting Water Use

**Time of day and day of week.**    In most communities, water use is lowest in the early morning hours, when most people are asleep. On a typical day, use rises until midday, stabilizes during the afternoon, and peaks during the early evening hours. The changing hourly rate of water use for a typical day is shown in Figure 22. Different days of the week show different total water uses; the day-to-day pattern depends on the habits of the community, but within a season the pattern is usually repeated week after week with only minor variations.

**Climate and season of the year.**    Water use is usually highest during summer months, particularly in warm, dry climates. More water is used for lawns and gardens, bathing is more frequent (also true in excessively humid areas), and air coolers that cool by evaporation of water ("swamp coolers") are more widely used.

During winter months in cold climates, water use may be surprisingly high. In some areas consumers run water faucets continuously to prevent water from freezing and bursting the pipes. Some water systems follow the same practice to protect water mains above the frost line.

**Type of community.**    Residential communities will use less water per person than highly commercialized or industrialized communities. The type of housing that is most common will also affect use. Low density residential areas (those with few housing units per acre) with large gardens and lawns will have a higher water use per person than higher density areas with multiple family dwellings such as townhomes, condominiums, and apartment complexes.

**Household water use.**    Various household appliances, such as dishwashers and garbage disposals, and facilities such as additional bathrooms, all once considered luxury items, are now common in the US. Each requires water, and this is causing water use per person to increase. Drought conditions may force many communities to impose mandatory restrictions on water use, and records of water use prepared and maintained by the operator are valuable in showing consumers how to conserve water during such times.

**Water pressure.**    As water pressure increases, the amount of water the consumer uses increases. A pressure of 25 to 50 pounds per square inch gage (psig) is considered normal. By increasing a 25 psig service pressure to 45 psig, water use can be expected to increase as much as 30 percent.

**Metering.**    Converting a flat-rate, non-metered system to a metered system has been shown to reduce water use by as much as 25 percent. Routine inspection

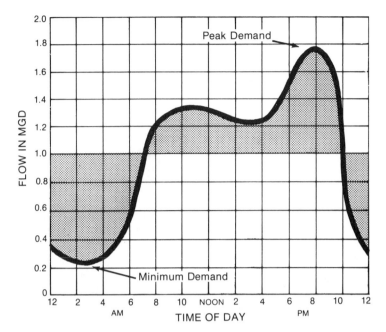

**Figure 22.  Typical Daily Flow Chart Showing Peak and Minimum Demands**

and maintenance of water meters is necessary to ensure that they continue to record the consumer's water use accurately.

**Air conditioning.**  Home air coolers that operate by evaporating water ("swamp coolers") are being replaced with air conditioners that cool by refrigeration methods using no water. However, in large commercial and industrial facilities, water-evaporation air coolers are still widely used. The demand for water to support air coolers is seasonal and is affected by daily temperature variations.

**Water quality.**  People will use less public water if it has an unpleasant taste, odor, or color or if it is hard or highly mineralized. This is a potentially dangerous situation, because the consumer may substitute water from a pleasant-tasting, but unsafe, private supply for the unpleasant tasting, although safe, water from a public supply.

**Sewers.**  The availability of municipal sewer systems increases water use. Homes with sewer service usually have more appliances and fixtures that use water (such as garbage disposals and extra baths) because those homeowners need not be concerned with the treatment and holding capacities of a private septic system. The increase in water use due to sewer availability may be 50 to 100 percent.

**Condition of the water system.**  A change in the condition of the distribution system can greatly alter the amount of water a treatment plant must supply, while the volume of water actually used by consumers remains unchanged.

When *water mains* are installed they are tested for water tightness. A small amount of leakage is permitted because it is practically impossible to construct a completely water-tight line. A new distribution system, with a total of 100 miles of 18-inch diameter pipe, constructed to current standards of allowable leakage, would leak about 90,000 gallons per day (90,000 gpd). As the system ages, leakage increases. Improper bedding of the pipe during installation or settling of the trench may partially open joints. Hastily or improperly installed service connections can leak. Thrust in the pipe, caused by the forces of flowing water, tends to separate joints.[13] Joint compounds can deteriorate and develop leaks. Corrosion of pipe materials can cause leakage. All of these factors, and others, increase water-main leakage. Unless periodic checks are made to locate and repair leaks, *the amount of water lost daily can equal or exceed the amount used by consumers.* Consumers do not use or pay for lost water, so leakage of treated water can cause the utility major revenue losses.

Distribution system *storage tanks* are usually equipped with overflow outlets to prevent accidental over-filling. Tanks are also usually equipped with a valve called an altitude valve (because it is controlled by the height, or altitude, of water in the tank) that automatically stops the incoming water when the tank is full. Improperly operated or inadequately maintained altitude valves can stick open, causing the tank to overflow continuously. Sometimes the altitude valves are missing, either because they were never installed or because they were removed and not replaced. Jammed or missing altitude valves can cause hundreds of gallons of water to be wasted every minute.

Certain other types of *valves* in the distribution system (air relief, vacuum relief, and pressure relief valves[14]) can cause significant water loss. Vandals tampering with fire hydrants can also cause a large water loss problem.

## Volume and Flow Rate

The operator must make measurements and maintain accurate records of the volume and rate of water passing through the water system. Operators use these records in planning daily operation and identifying operating problems; managers use the records for such things as planning system expansion, determining water-use charges, and planning water conservation measures. Two types of water measurement are commonly made in treatment plants: volume and flow rate.

**Volume.** VOLUME is simply the measurement of an amount of water; it is expressed in gallons, millions of gallons, or cubic feet. A water volume measurement may describe the amount of water contained in a reservoir; or the measurement may describe the amount of water that has passed by or through a certain point, such as a treatment plant.

**Flow rate.** When you measure both the *volume* of water passing by a point and the *time* it takes the water to pass, it gives you a measurement of FLOW RATE. Flow rate measurements are either *instantaneous* or *average.*

[13] *Basic Science Concepts and Applications,* Hydraulics Section, Thrust Blocks.
[14] *Basic Science Concepts and Applications,* Hydraulics Section, Flow Rate Problems.

INSTANTANEOUS FLOW RATE describes the rate at which water is passing by a point at any instant. It is defined[15] as

$$Q = AV$$

where
$Q$ = instantaneous flow rate
$A$ = the cross sectional area of the moving water
$V$ = the velocity (speed) of the moving water

The instantaneous flow rate varies constantly throughout the day. For example, Saturday at 6 PM the flow rate might be 6.3 million gallons per day (mgd); at 7 PM the same day it might be 5.8 mgd. It is important to understand that these are flow *rates*, and not flow *volumes*. If the water continued to flow at a constant 5.8 mgd flow rate for 24 hours, only then would the volume for the day be 5.8 million gallons.

Average flow rate for a period of time can be found by averaging the instantaneous flow rates during that period. The average flow rate can also be calculated by dividing the total volume of water by the length of time it takes the water to pass. For example, if 4.8 million gallons passed through the plant in 2 days, then the average flow rate for the 2-day period was 2.4 mgd (4,800,000 gallons ÷ 2 days).

You may hear the term FLOW used to mean instantaneous flow rate, average flow rate, or even volume; however, if you understand the concepts of flow rate and volume, then you will usually know what is being referred to. For example, if someone says, "The flow at 3 PM was 1.7 mgd," you will know they are talking about instantaneous flow rate, because they have given a specific instant when the measurement was made. However, if someone says, "The flow for Monday was 1.7 mgd," then, because they have specified a definite time period and units of volume per time, you will know they are talking about average flow rate. And if someone says, "The total flow for 1978 was 755 million gallons," then you will know that they are referring to volume, because the measurement is in gallons.

## Metering

In most treatment plants, the volume and instantaneous flow rate of water passing through the plant are continuously recorded by automatic metering systems; Figure 23 shows one type of recorder used with these systems.

Volume is read from the *totalizer*, the row of numbers on the lower left side of the recorder shown in the figure. These numbers, which look like a car's odometer, show the total volume of water that has passed through the plant at the time the reading is taken.

Flow rate can be determined by reading the totalizer, waiting a definite period of time, and then reading the totalizer again—subtracting the first reading from the second gives the volume that passed during that time. For example: Sunday morning at 8 AM the operator records a totalizer reading of 5,000,000 gallons. Monday morning at 8 AM (24 hours later) the operator records a totalizer

---

[15] *Basic Science Concepts and Applications,* Hydraulics Section, Flow Rate Problems.

reading of 6,500,000 gallons. Subtracting, the operator finds that a volume of 1,500,000 gallons passed through the plant during the 24-hour period; so the flow rate for that day was 1,500,000 gallons per day (1.5 mgd).

Instantaneous flow rate is read, at any moment, from the speedometer-like indicator at the top of the recorder. Instantaneous flow rate is also recorded continuously on the circular chart in the large central window of the recorder. A circular chart, made by a recorder like the one in Figure 23 and showing how instantaneous flow rate varied for one week, is shown in Figure 24. Notice how the instantaneous flow rate was always changing—this is often the case, especially with plants in smaller systems. For instance, at 9 PM Friday the chart shows that the instantaneous flow rate was 4000 gpm; at 3 AM Tuesday it was 1600 gpm.

## Readings and Calculations

The readings and calculations of water use that you will have to make most often are explained in the following paragraphs. The terms of measurement are summarized at the end of the section on Table 11.

**Instantaneous flow rate.**    This measurement, explained in the preceding section, determines the amount of treatment chemicals that should be added to the water being treated at any instant. Most treatment plants are equipped with systems that automatically vary the amount of chemical additions to match the changing flow rate.

*Courtesy of Muesco, Inc.*

**Figure 23.    Typical Circular Chart Recorder**

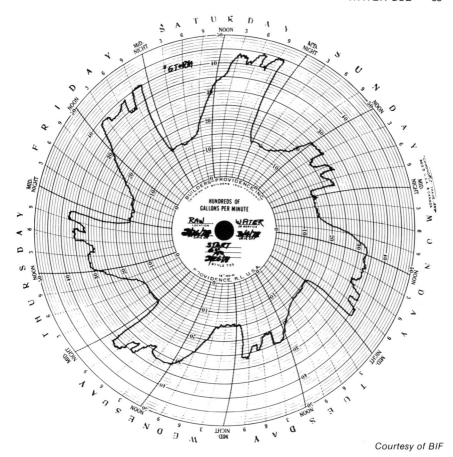

*Courtesy of BIF*

**Figure 24.   Typical Circular Chart Recording**

**Daily flow and average daily flow.**    The total number of gallons of water that passes through the plant during a 24-hour period is the DAILY FLOW. It is expressed as gallons per day (gpd) or million gallons per day (mgd). To find the AVERAGE DAILY FLOW (ADF) find the sum of all daily flows for a specified time period and divide by the number of daily flows used.[16] Records of daily flows and average daily flows are the basic information from which many other important calculations are made.

**Annual average daily flow.**    For each year, the ANNUAL AVERAGE DAILY FLOW is computed by averaging the daily flows for the year—that is, by adding them together and dividing by 365, the number of daily flows added. For example, if the sum of the daily flows for 1978 is 730 million gallons, then the plant's annual average daily flow for 1978 would be

$$\frac{730 \text{ million gallons}}{365 \text{ days}} = 2 \text{ mgd annual average daily flow}$$

---

[16] *Basic Science Concepts and Applications,* Mathematics Section, Average Daily Flow.

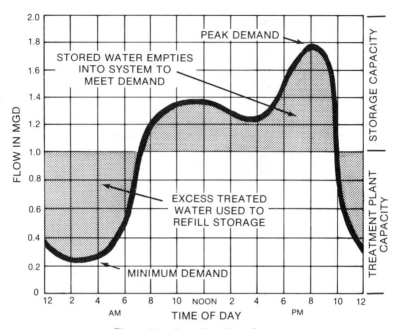

**Figure 25.  Peak Day Flow Curve**

This is an average; the daily flow for some days was less than 2 mgd, and the daily flow for other days was more. The annual average daily flow is important in system expansion; when annual average daily flow has increased to 70 percent of the plant's capacity, planning for new facilities should be underway.

**Peak day demand.**    During one day of each year, usually in midsummer, the plant will deliver more water than on any other day of the year. The amount of water delivered for that day is the PEAK DAY DEMAND, and the day on which the peak day demand occurs is the PEAK DAY. Figure 25 shows how the flow rate changed during one year's peak day and what the operator did to meet the demand: During the hours between midnight and 7 AM and between 10 PM and midnight, the flow being used by the system's consumers was less than the treatment plant capacity; however, the operator continued to operate the plant at full capacity through the night, allowing the treated water not used by consumers to go into storage. Between 7 AM and 10 PM, the consumers demanded more than the plant could deliver; however, the operator was able to supply the demand by supplementing the plant's output with the water already treated and in storage. Knowing when in the year the peak and other high-demand days will occur is important to the operator, who must make sure that the system is in top operating condition and that each night the system storage tanks are completely filled.

The peak demand and how it compares with average daily flow rate is also important to management. Plants are commonly built with design capacity equal to the peak day demand; as population grows and peak day demand increases beyond plant design capacity, water rationing or construction of new facilities must be considered.

**Minimum day demand.**  During one day of each year, usually in the winter, the plant will deliver less water than any other day of the year. The amount of water delivered for that day is the MINIMUM DAY DEMAND, and the day itself is the MINIMUM DAY. Knowing approximately what time of year the minimum day demand will occur is important to the operator. The output of the plant may need to be reduced, and portions of the plant may even need to be shut down for part of the day to avoid treating more water than can be used or stored.

**Peak month demand.**  During one calendar month of each year, usually during the summer, the plant will deliver more water than during any other month of the year. The volume of water used during that month is the PEAK MONTH DEMAND; the month itself is the PEAK MONTH. Operators and management need to know when to expect the peak month, and approximately how much water will be required, so they can be sure that enough raw water is in storage to supply the system's needs.

**Minimum month demand.**  During one calendar month of the year, usually in the winter, the plant will deliver less water than during any other month of the year. The volume of water used during that month is the MINIMUM MONTH DEMAND; the month itself is the MINIMUM MONTH. It is important for the operator to know when the minimum month will occur, since this is the best time to schedule major maintenance, repair, and replacement.

**Other measurements.**  The greatest and least amounts of water a system delivers for a full hour are the PEAK HOUR DEMAND and the MINIMUM HOUR DEMAND. They are found by examining the chart recordings showing hourly flow rates. These measurements, made by the operator, are important to engineers designing new distribution systems. Partly based on current hourly demand figures, the engineer can predict what the water needs of the community will be in the future. He then engineers a system with a design capacity equal to the future need.

A DEMAND FACTOR is the ratio of a peak or minimum demand to the average demand. For instance, if the peak day demand is 6 mgd, and the average daily flow rate is 3 mgd, then the peak is twice the average, and the peak day demand factor is 2. The larger the community, the less the peaks and minimums vary from the average. Demand factors are used in system design.

**Table 11.  Common Measurements of Water Use**

| Term | Definition | How to Calculate | Units of Measure | How used / Comments |
|---|---|---|---|---|
| Volume | Amount of water which flows through plant | Read totalizer at beginning and end of period for which volume is to be determined, and subtract first reading from second | gallons, millions of gallons, cubic feet, acre feet | One of the basic measurements of water use |
| Instantaneous Flow Rate | The flow rate at any instant — defined by the equation $Q = AV$ | Read directly from circular flow chart or from indicator<br><br>*or*<br><br>Use $Q = AV$ equation | gallons per minute (gpm), gallons per hour (gph), gallons per day (gpd), million gallons per day (mgd) cubic feet per second (cfs) | One of the basic measurements of water use |
| Daily Flow | Amount of water passing through plant during a single day | Measure flow volume (gallons) for a single day | gpd, mgd | Basis for calculating ADF and chemical dosages |
| Average Daily Flow | Average of daily flows for a specified time period | Find sum of all daily flows for the period and divide by number of daily flows used | gpm, gph, gpd, mgd | Records of average daily flow rate are the basis for calculating most other important measurements |
| Annual Average Daily Flow | Average of average daily flows for a 12-month period | Average the average daily flows for all the days of the year<br><br>*or*<br><br>Divide total flow volume for the year by 365 | gpm, gph, gpd, mgd | Used to predict need for system expansion |

| Term | Definition | How Determined | Units | Significance |
|---|---|---|---|---|
| Peak Day Demand | Greatest volume per day flowing through the plant for any day of the year | Look at records of daily flow rates for the year to find the peak day | gpm, gph, gpd, mgd | Determines system operation (plant output, storage) during heaviest load period—ranges from 1.5 to 3.5 times the average daily flow rate |
| Minimum Day Demand | Least volume per day flowing through the plant for any day of the year | Look at records of daily flow rates for the year to find the minimum day | gmp, gph, gpd, mgd | Determines possible plant shut-down periods—ranges from 0.5 to 0.8 times the average daily flow rate |
| Peak Month Demand | Greatest volume of water passing through plant during a calendar month | Look at records of monthly flow volumes (determined from totalizer readings at the beginning and end of each month or from the sums of the daily flows for each month) to find peak month for the year | gallons, millions of gallons, billions of gallons | Helps determine raw water storage needs — peak month demands range from 1.1 to 1.5 times the average monthly flow volume for the year. |
| Minimum Month Demand | Least volume of water passing through plant during a calendar month | Look at records of monthly flow volumes (determined as for Peak Month Demand) to find minimum month for the year | gallons, millions of gallons, billions of gallons | Best time of year for checking and repairing equipment — minimum month demands range from 0.75 to 0.90 times the average monthly flow volume for a year. |
| Peak Hour Demand | Greatest volume per hour flowing through the plant for any hour in the year | Determined from the chart recordings showing the continuous changes in flow rate | gpm, gph, gpd, mgd | Determines required capacity of distribution system piping — peak hour demands range from 2.0 to 7.0 times the average hourly demand for the year. |
| Minimum Hour Demand | Least volume per hour flowing through the plant for any hour in the year | Determined from the chart recordings showing the continuous changes in flow rate | gpm, gph, gpd, mgd | Minimum hour demands range from 0.20 to 0.75 times the average hourly demand for a year. |

## Selected Supplementary Readings

*Manual of Instruction for Water Treatment Plant Operators.* New York State Dept. of Health (no date). pp. 15 - 16.

*Estimated Use of Water in the U.S. in 1975.* Geological Survey Circular 765 (1977). pp. 3 - 7, 10 - 13, tables 5 and 12.

*A Primer on Water Quality.* Geological Survey, US Dept. of Interior (1965). pp. 22 - 23.

*A Primer on Water.* Geological Survey, US Dept. of Interior (1960). pp. 32 - 34.

*Water Facts and Figures for Planners and Managers.* Geological Survey Circular 601 - 1 (1973). pp. 8 - 13.

*A Customer Handbook on Water-Saving and Wastewater-Reduction.* The Washington Suburban Sanitary Commission (rev. April 1975).

Babbitt, Harold E.; Doland, James J.; & Cleasby, John L. *Water Supply Engineering.* McGraw-Hill Book Company (1962). pp. 10 - 14, 14 - 18.

## Glossary Terms Introduced in Module 2

(Terms are defined in the Glossary at the back of the book.)

| | |
|---|---|
| Annual average daily flow | Minimum day demand |
| Average daily flow | Minimum hour demand |
| Commercial water use | Minimum month |
| Daily flow | Minimum month demand |
| Demand factor | Peak day |
| Domestic water use | Peak day demand |
| Flow | Peak hour demand |
| Flow rate | Peak month |
| Gallons per capita per day (gpcd) | Peak month demand |
| Industrial water use | Public water use |
| Instantaneous flow rate | Volume |
| Minimum day | |

## Review Questions

(Answers to Review Questions are given at the back of the book.)

1. What are the four principal water uses?

2. What was the national average daily water use per person for 1975? What amount was used domestically?

3. In 1977 a small community in Georgia delivered drinking water to the consumers at an average of 0.63 mgd. The water system served a population

of 2250. What is the average flow rate expressed in gallons per capita per day (gpcd)?

4. What are the four major categories of water use from public water systems?

5. (a) Typical domestic water uses and the amount each consumes are listed in the text by household fixture or type of use. Calculate the average amount of water used each day by a four-member household that puts water to the following uses. Where a range of water use is given (such as laundering = 20 to 45 gallons per load), use the higher value (45 gallons). Express the answer in gallons per capita day (gpcd).

> Laundering ........................ 6 loads per week
> Shower............................ 3 per day
> Tub................................ 1 per day
> Dishwashing ....................... 2 loads per day
> Toilet ............................ 14 flushes per day
> Garbage disposal................... 2 minutes per day
> Lawn watering ..................... 4500 ft$^2$
> Car washing ....................... 10 minutes per week
> Drinking .......................... (family of four)

(b) Identify the one use which constitutes the highest water use per day.

6. The following flow rates were read from a circular chart recording for one 24-hour period at the plant. Flows are in mgd.

| Time | Flow | Time | Flow | Time | Flow |
|------|------|------|------|------|------|
| 8 AM | 1.20 | 4 PM | 1.25 | 12 MID. | 0.35 |
| 9 AM | 1.30 | 5 PM | 1.35 | 1 AM | 0.25 |
| 10 AM | 1.35 | 6 PM | 1.50 | 2 AM | 0.20 |
| 11 AM | 1.35 | 7 PM | 1.70 | 3 AM | 0.20 |
| 12 Noon | 1.30 | 8 PM | 1.75 | 4 AM | 0.25 |
| 1 PM | 1.25 | 9 PM | 1.70 | 5 AM | 0.40 |
| 2 PM | 1.20 | 10 PM | 1.10 | 6 AM | 0.55 |
| 3 PM | 1.20 | 11 PM | 0.50 | 7 AM | 1.00 |

(a) Identify the peak hour demand.

(b) Identify the minimum hour demand.

7. Which of the following are considered commercial uses of water?

   (a) canneries    (c) textile mills   (e) public parks

   (b) car washes   (d) laundries

8. A new cannery is being planned for construction in the area and they have requested that all their water needs be met by water from your treatment plant. At full production this cannery will process 30 tons of peaches per day. How much water will the cannery require daily?

9. Define the term "peak day demand." From what period of records are peak day demands determined?

10. List five factors that affect water use, and describe the effect of each.

## Study Problems and Exercises

1. Based on the percentages given in Figure 21 and using Table 8, calculate the amount of water used daily in your state for domestic, commercial, industrial and public uses. Report your answers as gpcd.

2. Select one industry served by your water system. Based on water use for the most recent 12 months of record, prepare a brief report on its industrial water use, identifying

   (a) total annual use in gallons

   (b) monthly average, maximum and minimum use

   (c) water used per unit of product produced

   (d) water cost associated with each unit of product produced.

3. Based on flow meter records at your water supply system, determine the total water delivered by the treatment plant for the most recent 12-month period. Using a current estimate of population served, report your answer as gpcd.

4. Based on the same data used in problem 3, calculate for your water system

   (a) peak day flow, in gpd, noting day of week and date

   (b) peak month volume, noting month and year

   (c) minimum day flow, in gpd, noting day of week and date

5. Using data from the same one-year period as in problem 4, plot the 24-hour water use curve for one day in the month of June or July and one day (the same day of the week) in December, January or February. (See example curve Figure 22 in text.) Compare the results, identify the peaks, and explain the reasons for each peak.

6. Using the results of problems 4 and 5 above compare actual water use to water system design capacity, checking

(a) transmission capacity for raw water (Note: transmission lines are usually designed to handle the peak day demand)

(b) treatment capacity (Note: treatment plants are usually designed to handle the peak day demand)

7. From the results of problem 5 answer the following questions:

(a) Are winter flows excessive?

(b) Are minimum flows excessive, suggesting leaks or continuous flows to storage?

Check your results against wastewater treatment plant flow data. Consumer taps allowed to run to prevent freezing will usually register at the wastewater plant as unexplained high flows during the early morning period, between 12 midnight and 6 am. Excessive water flows due to water system leaks will usually not show in the wastewater flow records (unless by sewer infiltration). When checking wastewater flow records, particularly on older systems, consider that ground-water infiltration may be a contributing factor to early morning high flows, although not usually during the winter.

8. How many consumers in your water system receive unmetered water service? Assuming that meters would reduce water use by 25 percent, how much water could your community save by installing meters?

9. Compare the total volume in a given month as recorded by the water system master meter against the total of all volumes billed that month to consumers. Explain any differences in these two totals.

10. From a strip or circular chart showing the flow rate of water entering a treatment plant during a one-day period, determine the following:

(a) What was the peak instantaneous flow rate and when did it occur?

(b) What was the minimum instantaneous flow rate and when did it occur?

(c) During what time period would it have been best to perform routine maintenance and repair for the day?

# Module 3

# Developing the Water Supply

Developing a water supply is a complex task requiring the expertise of engineers, geologists, and other scientists. However, after the supply is developed it is the operator's responsibility to maintain and operate the system. To meet this responsibility you need a basic understanding of the raw water supply system—what it is and how it operates.

After completing this module you should be able to

- Describe the three primary types of surface storage impoundments.
- Describe the important components of impounding reservoirs.
- Describe four common problems associated with operating and maintaining surface storage impoundments.
- Define the terms and measurements used to describe water wells.
- Name several methods of constructing a well, and describe the components of wells constructed in loose materials.
- State the sanitary conditions needed for a safe ground-water supply.

## 3-1. Surface Water Development

A water supply must be reliable. It must provide all the water the community requires, every day of the year. The flow in any water course changes with rainfall and runoff—the amount that can be counted on for diversion to the water system can never be more than the natural flow during the low-flow season or during drought conditions. Very often this low-flow condition is not adequate to meet the community's needs.

Community water needs are normally greatest during the warm months of summer and early fall. But this is the period when natural stream flow is the lowest. To resolve this problem and increase water supply reliability, impounding reservoirs are constructed. The IMPOUNDMENTS store water during periods of excess supply so that it can be used during periods when natural stream flow is deficient.

## Types of Storage Impoundments

**Natural storage impoundments.**   *Lakes and ponds* that offer an adequate supply of quality water make excellent natural storage impoundments. They should be located close to the treatment plant and the community—otherwise lengthy and costly aqueducts and pipelines must be constructed. The impoundment should be located in a watershed area with little or no development by man, or it must be protected from pollution caused by agricultural runoff, wastewater discharges, feed-lot runoff, and other man-made pollution problems or sources.

**Man-made storage impoundments.**   Until the mid-1800's, most man-made storage impoundments were *excavated,* made by digging out a shallow basin and using the excavated material to build up embankments. The cost of excavated impoundments usually limits their use to small capacity situations. An excavated impoundment is shown in Figure 26.

With the development of modern construction equipment and techniques, dams have become a common way of creating a reservoir. They are usually built where a large reservoir can be formed by damming a deep and narrow valley. Many different materials and methods of construction are available. The most common dams are

- Embankment dams—constructed of earth, rock, or both.
- Masonry dams—constructed of concrete.

**Recharged ground-water aquifers.**   Recently, ground-water aquifers have been used to store water from surface sources. Surface water is pumped (injected) into low or depleted aquifers during high surface flow periods, and the water is pumped back out of the aquifer when needed. Although this achieves a high-

**Figure 26.   Excavated Impoundment**

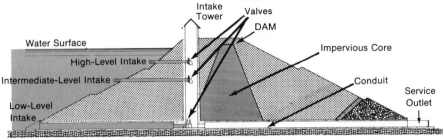

**Figure 27. Dam and Intake Tower for Impounded Surface Water Supply**

Reprinted with permission of John Wiley & Sons, Inc., New York, N.Y., from *Elements of Water Supply and Wastewater Disposal*, 2nd ed., by Gordon Fair, John Geyer, and Daniel Okun. Copyright ©1971.

volume storage capacity at low cost, the surface water can pollute the aquifer. Therefore water from this type of storage must be properly treated prior to use. Another problem is that the volume of water pumped into the aquifer cannot always be fully recovered.

## Important Components of Impounding Reservoirs

**Intake structures.** Any time surface water is impounded, an INTAKE STRUCTURE is needed to withdraw the water. This is the component of impounding reservoirs with which you will be most concerned because you may be involved in its operation. Intake structures range from simple pipe or overflow outlets to complex towers (see Module 4). Multiple-level intakes, shown in Figure 27, allow the operator to withdraw water from various depths to compensate for a changing water level, to select the best depth from a water quality standpoint, and to avoid an ice cover.

**Spillways.** The capacity of an impounding reservoir may be too small to contain all the water that flows into it, particularly during heavy flood conditions. A spillway is the part of a dam that allows excess water from the reservoir to flow into the stream or river below the dam. In some cases, two spillways are constructed: (1) a service spillway passing flows of low to moderate volumes during normal operation; (2) an emergency spillway used almost exclusively during extremely high flow conditions. Figure 28 shows a typical chute spillway of a small concrete dam.

**Service outlets.** The service outlet allows water to be released when needed by users located downstream from the impounding reservoir. It also may be used to lower the water level in the reservoir. The outlet is designed to release water regardless of the water level in the impoundment (unlike the spillway, which can only pass water when the level is high). The service outlet usually includes a conduit and gates or valves located near the bottom of the reservoir (Figure 27).

## Operating Problems of Impoundments

Some common problems with the operation of impoundments are

- Evaporation and seepage
- Silting
- Maintaining good water quality

**Figure 28.   Typical Chute Spillway on a Concrete Dam**

**Evaporation and seepage.** Storage impoundments with large exposed surface areas commonly lose from 6 to 8 ft of water a year through evaporation. In addition, there are SEEPAGE losses through the bottom and sides of the impoundment. In large reservoirs these losses cannot be economically controlled; therefore the reservoirs are constructed with a capacity greater than the volume that will be withdrawn, so that even after evaporation and seepage enough water remains in storage to provide an adequate supply of good quality water. In small reservoirs, particularly small excavated reservoirs, seepage can be controlled by lining the impoundment, as shown in Figure 29. Evaporation can be controlled by covering the impoundment. Covers range from reinforced concrete slabs (on which some communities have constructed recreational facilities such as tennis courts), to floating plastic sheets (Figure 30), to thin layers of a liquid chemical. Covering also protects water quality.

**Silting.** The raw water entering an impoundment often carries silt (finely ground rock, much finer than sand). The quiet, slow moving conditions of an impoundment allow this silt to settle. In time, silt can fill a significant portion of the impoundment's storage capacity. SILTING can be a major problem with all surface storage facilities, and there is no simple solution. The amount of silt reaching the impoundment can be controlled through good watershed management methods, such as proper planting on the watershed and good farming practices. Once an impoundment becomes silted, however, only draining the impoundment and excavating the sediments can restore it to full capacity. This is usually not an economical solution for major reservoirs, so reservoirs are often designed and constructed with extra capacity to allow for silt accumulation.

**Maintaining a good raw water quality.** Maintaining good raw water quality will be of prime concern to you, and it is an area where your efforts can return measureable benefits to the community you serve. Taste and odor problems and the problems due to algae blooms are the main quality difficulties associated with storage impoundments. Detailed information on maintaining or improving raw-water quality is discussed under the topic of preliminary treatment in Volume 2, Module 1.

## 3-2. Ground-Water Development

Most small water supply systems in the US take their raw water supply from ground-water sources, and operators of small systems are often involved in the development of ground-water supplies. To a lesser extent, operators of large systems also may be involved in ground-water development.

### Exploring for Ground Water

Ground water can be found just about anywhere under the earth's surface. However, for use as a public water supply the ground water must be of drinking quality. Water must move through the aquifer to the well fast enough to be

*Courtesy of Burke Rubber Company, a Division of Burke Industries*

**Figure 29. Lining Small Impounding Reservoir With Plastic**

*Courtesy of Burke Rubber Company, a Division of Burke Industries*

**Figure 30. Plastic Covered Reservoir**

pumped at the desired flow rate on a sustained basis. And the aquifer should be within economical drilling distance of the surface.

Ground-water development, like surface water development, is a job for engineers and scientists experienced in hydrology, geology, and hydraulics. Experts should always be consulted on large-scale projects because of great capital investment risk and the public's dependence on an adequate supply of safe drinking water. Often the operator can greatly assist the exploration effort by having a basic understanding of what is involved.

**Existing data.**    Before going into the field, the geologist will want to review geologic data, maps, and the logs of your existing wells. A typical well log (Figure 31) shows the geologic formations in which the well is drilled. By examining several well logs the geologist gains an idea of the geology of the area, which helps him locate possible productive aquifers. This background information is often available through state water resources agencies. Sometimes existing information is all that is needed to complete the exploration; in other cases more information is required and the geologist must perform a field investigation.

**Field investigations.**    During the field investigation, the geologist will take a closer look at any surface evidence of ground water occurrence. Among the features that would produce valuable clues would be land forms, stream patterns, springs, lakes, and vegetation.

Ground water is likely to be present in larger quantities under valleys than under hills. Valley soils containing PERMEABLE material washed down from mountains are often productive aquifers. Coastal terraces as well as coastal and river plains also may indicate good aquifers.

Any evidence of surface water, such as streams, springs, seeps, swamps, or lakes, is a good indication that there is ground water present, though not necessarily in usable quantity. The sand and gravel deposits found in river beds may very often extend laterally (to the sides) into the river banks and beyond. If they do, then a shallow well in that area can be highly productive.

Sometimes vegetation will show the location of ground water. An unusually thick overgrowth may indicate shallow ground water.

The next step is physical testing. Usually the first tests are the seismic test and the resistivity test. The seismic test measures the speed that a shock wave travels through earth; it is used to help identify underground geologic formations that may contain ground water. The resistivity test measures the ground's electrical resistance; generally, the less the electrical resistance, the greater the probability of water. Both tests are made at the ground surface—no drilling is needed.

Next, exploration holes are drilled in key locations and the holes are logged electrically or by gamma-ray. The electric log is a measure of the change in the earth's resistance to an electric current as depth increases; it is used to supplement the descriptive logging of the hole, which the driller makes based on samples taken at increasing depths as drilling proceeds. Gamma-ray logging—a measurement of how controlled radiation penetrates the earth—helps to identify the various formations encountered in the exploration holes.

For a large project, test holes may be drilled and used with previously drilled

ωA   RECEIVED

COLORADO DIVISION OF WATER RESOURCES
1313 Sherman Street - Room 818
Denver, Colorado 80203

MAR 09 '78

THIS FORM MUST BE SUBMITTED
WITHIN 60 DAYS OF COMPLETION
OF THE WORK DESCRIBED HERE-
ON, TYPE OR PRINT IN BLACK
INK.

WELL COMPLETION AND PUMP INSTALLATION REPORT
PERMIT NUMBER ___R16368-RF___

WATER RESOURCES
STATE ENGINEER
COLO.

WELL OWNER ___Two H B Farms___

ADDRESS ___20908-Co. Rd. 28.5, Brush, Co. 80723___

DATE COMPLETED _____January 13___, 19 78

___SW___ ¼ of the ___SE___ ¼ of Sec. 23 ___,

T. __4__ N , R. __56__ W , __6th__ P.M.

HOLE DIAMETER

___34___ in. from ___0___ to ___103½___ ft.

WELL LOG

| From | To | Type and Color of Material | Water Loc. |
|------|-----|------|------|
| 0 | 3 | Top Soil | |
| 3 | 7 | loam | |
| 7 | 35 | coarse sand | 23' |
| 35 | 70 | watersand, fine gravel | |
| 70 | 84 | gravel, small rocks | |
| 84 | 84½ | clay | |
| 84½ | 103½ | gravel, small rocks | |

_____ in. from _____ to _____ ft.

_____ in. from _____ to _____ ft.

DRILLING METHOD_____

CASING RECORD:    Plain Casing

Size __16"__ & kind Transite from __± 1½__ to __38.5__ ft.

Size _____ & kind _____ from _____ to _____ ft.

Size _____ & kind _____ from _____ to _____ ft.

Perforated Casing

Size __16"__ & kind Transite from __38.5__ to __103½__ ft.

Size _____ & kind _____ from _____ to _____ ft.

Size _____ & kind _____ from _____ to _____ ft.

GROUTING RECORD

Material _____Cement_____

Intervals _____0-10_____

Placement Method ___Spill Tube___

GRAVEL PACK:  Size Pea & Buckshot, mixed 10-70

Interval _____Pea_____  70-103½

TEST DATA

Date Tested _____January 14___, 19 78

Static Water Level Prior to Test _23_____ ft.

Type of Test Pump __8" Vertical Turbine__

Length of Test _____5 hrs._____

Sustained Yield (Metered) _____2050 gpm

TOTAL DEPTH __103½'__

Use additional pages necessary to complete log.

Final Pumping Water Level ___34'-9"___

*Courtesy of State of Colorado, Division of Water Resources*

**Figure 31.  Typical Well Completion Log**

exploration holes for measuring an aquifer's production characteristics. Finally, background and field data are combined and used to select the well site.

## Water Well Construction

A general understanding of the construction of water wells will help you to operate and maintain them properly.

**Construction techniques.** Wells can be hand-dug, bored with earth auger, driven using a drive-well point, jetted using water under pressure, or drilled. To supply the capacities and achieve the level of sanitary protection required by public water systems, wells are usually drilled. The most common well-drilling techniques are (1) percussion drilling—repeated dropping of a bit against the material being drilled, and (2) rotary hydraulic drilling—cutting the material with a turning bit, and continuously removing cut-away material with a high pressure stream of water. The method used depends on the geology of the site and the availability of equipment.

**Parts of a well.** The parts of a typical well are labeled in Figure 32. At the surface, the SANITARY SEAL prevents contaminants from entering the well casing. The seal has openings into the well for the discharge pipe, pump controls (if the pump is submerged), and an air vent to let air into the casing as the water level drops. The WELL SLAB supports pumping equipment and aids in sealing against surface water. The well CASING is a tube, generally metal, placed in the well to keep the well open. GROUT is a mixture of cement, water, and sometimes sand, which is pumped into the ANNULUS, the space between the drilled hole and the casing. Grout seals the well against contamination. At the bottom of the casing, penetrating the aquifer, is the intake section of the well, which consists of a WELL SCREEN (INTAKE SCREEN) and a GRAVEL PACKING. The screen, a very important part of the well, provides enough open spaces for an unrestricted water flow, yet is strong enough to support the loose aquifer and fine enough to stop sand from entering the well. The gravel packing aids the screen in filtering sand and increases well efficiency; wells set in aquifers composed of coarser materials do not require artificial gravel packing.

## Water Well Terminology

To operate a ground-water system, you must understand some of the common terms used to describe ground water and wells.[17] See Figure 33.

The STATIC WATER LEVEL is the water level in a well when no water is being taken from the aquifer. It is normally measured as the distance from the ground surface to the water level in the well. This is an important measurement because it is the basis for monitoring changes in the water table.

When water is pumped out of the well at a constant rate, the water level in the well drops and then remains fairly stable at a level lower than the static water level. The water level in the well during pumping, measured from the ground to the water surface, is the PUMPING WATER LEVEL. This level determines the location of the intake or submerged pump.

The measure of the drop in the water level during pumping is called DRAWDOWN. It is the difference between the static water level and the pumping water level.

During pumping, the water table is drawn down in the area surrounding the well. However, beyond a certain distance, the water table is not affected. As

---

[17]*Basic Science Concepts and Applications,* Mathematics Section, Well Problems.

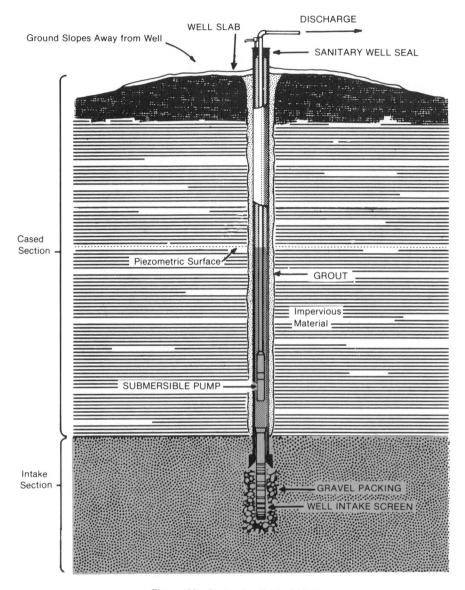

**Figure 32. Parts of a Typical Well**

shown in Figure 33, the RADIUS OF INFLUENCE is the distance from the well within which the water table is influenced or drawn down when the well is being pumped. You will also notice from Figure 33 that the water surface inside the radius of influence is shaped somewhat like a cone. This is called the CONE OF DEPRESSION.

After pumping is stopped the water level will rise in the well, toward the static

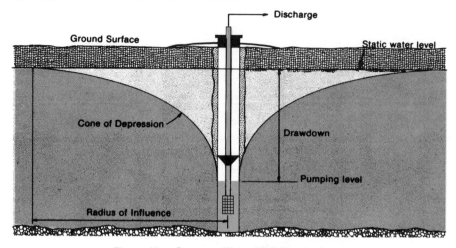

**Figure 33. Common Water Well Terminology**

water level. If the water level does not quite reach the original static water level, the distance it falls short is called RESIDUAL DRAWDOWN.

WELL YIELD is the rate of water withdrawal that the well can supply over a long period of time. Well yield is commonly measured as gallons per minute (gpm). For small yield wells the unit of measure can be gallons per hour (gph). For large yield wells cubic feet per second (cfs) is the unit of measure.

Perhaps the most important term in well testing and operation is SPECIFIC CAPACITY. Specific capacity is a measure of well yield per unit of drawdown and is usually expressed as gallons per minute per foot of drawdown. For example, if the well yield is 200 gpm and the drawdown is found to be 20 ft, the specific capacity is 200/20 or 10 gpm per ft of drawdown. A sudden drop in specific capacity indicates problems, which should be identified and corrected (such as a clogged well screen).

## Sanitary Considerations

**Sources of ground-water contamination.** Wastewater from human and animal populations can contaminate ground water with disease-causing bacteria and viruses. Agricultural runoff (carrying fertilizers, herbicides, and pesticides) and industrial wastes (carrying heavy metals, phenols, and exotic organic compounds) can also contaminate ground waters. Contaminants can be carried into the aquifer with percolating waters from the surface, or they can enter the aquifer more directly through poorly sealed operating wells or old, uncapped, abandoned wells. To avoid the problems of disease, taste, and odor that contamination can cause, wells must be properly (1) located and (2) constructed.

**Locating wells to avoid contamination.** Wells must be located a safe distance from potential contaminants. Because of differences in PERMEABILITY, underlying materials, and thickness of soils and because of the varying rates and directions of surface- and ground-water movement, there is no safe distance that applies in all cases. In some areas experience has shown 200 ft to be a safe distance from bacterial contamination for wells constructed in sand and gravel or in sandstone formation. However, in certain areas of the country, public health regulations prohibit construction of wells that are within 500 ft (more in some areas) of a contaminant source if the well would be in the flow path of the contaminant. Clearly each proposed well construction must be evaluated on its own merits, and well locations that provide the highest practical degree of protection should be established.

When a new well is put into operation, the direction of ground-water flow can be changed because of the well's cone of depression (Figure 34). Water can move into the well from directions that would have required an "uphill" movement within the aquifer before water was drawn from the well, and it can move at high rate of flow because of the steepness of the cone. Therefore, when locating a well the purity of the water in the existing aquifer is not the only consideration—the aquifer as altered by the well must be the determining factor.

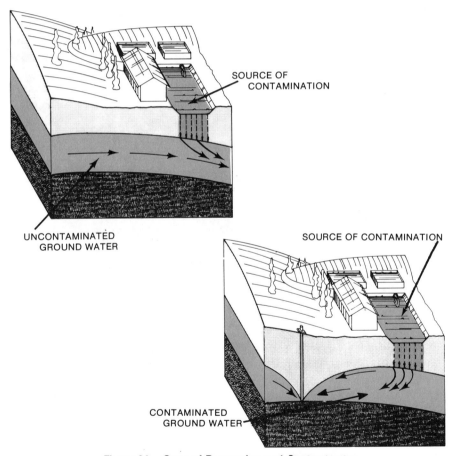

**Figure 34.  Cone of Depression and Contamination**
Operation of a well can draw contamination into a previously uncontaminated area.

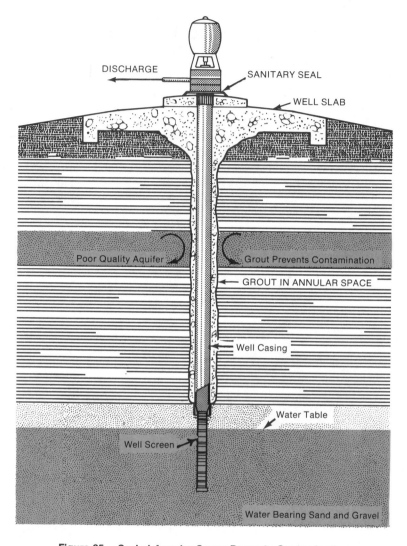

**Figure 35. Sealed Annular Space Prevents Contamination**

Wells should be located in formations sufficiently deep to protect against contamination by surface water percolation. The minimum depth needed to reach safe water varies widely, depending on the type of soil and rock formations. In some areas depths of 25 to 30 ft give reasonable protection in unconsolidated material. In other areas a minimum of 100 ft or more is required.

**Constructing wells to avoid contamination.** Proper well construction is one of the most important safeguards against contamination. Proper construction requires proper casing, grouting, and sealing.

Quality casings should be used. Your state may have casing standards to

AIR VENT

DISCHARGE LINE

PUMP POWER CABLE

SANITARY SEAL

WELL CASING

**Figure 36.   Sanitary Seal Detail**

follow. Thin-wall casing should be avoided because it may corrode rapidly or collapse under pressure. Used casing taken from abandoned wells should not be installed; it would be weak, and it might be contaminated, particularly if the abandoned well was used for something other than drinking water supply.

When wells are drilled, the hole is made much larger than the casing. The space between the casing and the hole, called the ANNULAR SPACE, or annulus, provides a direct channel from the surface to the ground water. This space must be filled with grout to prevent surface contaminants from running down the annulus and into the aquifer. The typical grout seal shown in figures 32 and 35 greatly minimizes the possibility of contamination.

Sealing the annulus has added advantages: It seals out poor quality water from an overlying aquifer (Figure 35); it increases the life of the well by protecting the casing against exterior corrosion; and it stabilizes the soil and rock formation to help prevent caving.

To complete well construction properly a well slab is placed and the sanitary seal is made (Figure 36). The sanitary seal plugs the top of the casing and seals around the well discharge pipe and air vent, preventing surface contaminants from directly entering the well. In Figure 35 note how the slab and ground should slope away from the well to prevent surface water from pooling around the well. Remember that the well slab is not a positive protection against contamination. Contaminants can get under the slab and into the well water if the well is not adequately grouted.

# Selected Supplementary Readings

### Surface Water Development

*Manual of Water Utility Operations.* Texas Water Utilities Association, (1975). pp. 40 - 43, 48 - 52.

### Ground-Water Development

*Manual of Individual Water Supply Systems.* Water Supply Division, EPA (reprint 1975). pp. 21 - 60.

*Ground Water.* Manual of Water Supply Practices, M21, AWWA (1973). pp. 14 - 23, 24 - 40.

Overman, Michael. *Water Solutions to the Problem of Supply and Demand.* Doubleday and Co., Inc. (1969). pp. 39 - 43, 78 - 81.

Gibson, Ulrich P. & Singer, Rexford D. *Water Well Manual.* Premier Press (1971). pp. 28 - 106.

*Ground Water and Wells.* Johnson Division, UPO Inc. (4th printing 1975). pp. 157 - 178, 206 - 208, 209 - 248, 295 - 312, 317 - 332, 395 - 412.


# Glossary Terms Introduced in Module 3

(Terms are defined in the Glossary at the back of the book.)

| | |
|---|---|
| Annular space (annulus) | Radius of influence |
| Casing | Residual drawdown |
| Cone of depression | Sanitary seal |
| Drawdown | Seepage |
| Gravel packing | Silting |
| Grout | Specific capacity |
| Impoundment | Static water level |
| Intake structure | Well screen (intake screen) |
| Permeability (permeable) | Well slab |
| Pumping water level | Well yield |


# Review Questions

(Answers to Review Questions are given at the back of the book.)

1. List the three major types of storage impoundments.

2. Name the structure used to withdraw water from a raw water supply for municipal use.

3. List the four major problems associated with surface water impoundments.

4. Define the term "static water level."

5. Is the following statement true or false? The pumping level is normally measured as the distance from the ground to the water surface in the well when pumping is in progress.

6. Measurements taken at a water table well produced the following information:

    Static water level........................................121 feet

    Pumping water level......................................141 feet

    Radius of influence......................................350 feet

    Well yield...............................................180 gpm

    Calculate the following:
    a) Drawdown.
    b) Distance from the well to the outer limit of the cone of depression.
    c) Specific capacity.

7. What is the public health significance of a well's cone of depression?

8. Identify the numbered items in the following figure.

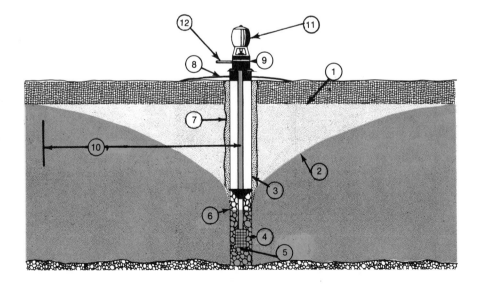

## Study Problems and Exercises

1. Identify the type of surface water source(s) used by your water supply system.

2. (a) Why is silting a problem in the operation of impoundments? (b) What can be done to control siltation? (c) Once deposited in an impoundment, how can silt be removed?

3. Describe the sanitary considerations necessary to ensure a safe ground-water supply.

4. Calculate the specific capacity of a well if drawdown is 12 feet and the well yield is 260 gpm.

# Module 4

# Water Transmission

WATER TRANSMISSION is the controlled movement of water from the source (a lake, river, reservoir, or well) to the treatment plant and from the plant to the WATER DISTRIBUTION system (Figure 37). The pipeline used to convey water in the transmission system is the TRANSMISSION LINE. The two typical water systems in Figures 38 and 39 show the general location of transmission lines.

Water transmission begins at the intake structure, which allows withdrawal of water from the source. The water then proceeds through aqueducts, pipelines, or both, to the treatment plant and from there to the distribution system. Depending on topography (changes in elevation), pumping may be required. The pipeline flow is controlled by an array of valves, and water movement is constantly measured with flow meters.

As an operator, it is your day-to-day responsibility to operate and maintain the transmission system. To do this, you must know the purpose and function of each of its components. In this module you will investigate the components of the transmission system.

After completing this module you should be able to

- Define water transmission

- Describe or identify the major components of a transmission system

- Identify the materials used in the manufacture of pipe for transmission systems

- Describe the types of valves used in transmission systems and explain the types of service performed by each

- Describe the two major types of pumps used in water transmission

- Discuss the importance of flow meters and describe how the types of meters used in transmission systems operate

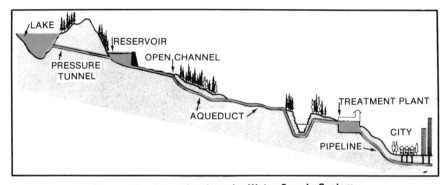

Figure 37.   Cross Section of a Water-Supply System

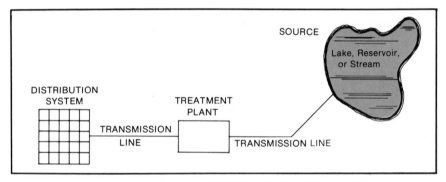

Figure 38.   Water Transmission From a Surface Water Source

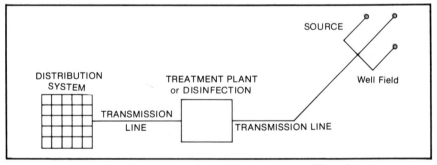

Figure 39.   Water Transmission From a Ground-Water Source

## 4-1.   Intake Structures

A transmission system begins at the water supply source with an INTAKE STRUC-
TURE. An intake structure is placed in a surface water source (such as a lake, river,
or reservoir) to withdraw water. Types of intake structures include simple
surface diversions, submerged intakes, movable intakes, pump intakes, and
infiltration galleries.

A.

*Courtesy of Fisher Scientific Company*

B.
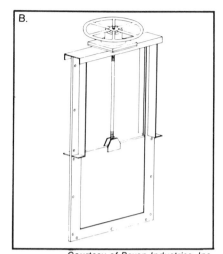
*Courtesy of Raven Industries, Inc.*

**Figure 40. Slide Gates**
A. Surface Diversion With Slide Gate   B. Hand-Wheel Operated Slide Gate

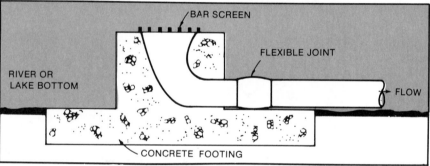

**Figure 41. Submerged Intake**
From *Water-Resources Engineering* by Ray K. Linsley and Joseph B. Franzini.
Copyright ©1964 by McGraw-Hill, Inc. Used with permission of McGraw-Hill Book Company.

*Surface diversions,* in their simplest form, consist of a small concrete structure containing a slide gate (Figure 40). When the slide gate is removed, water spills into a canal, ditch, or pipeline, which then carries the water to the treatment plant. This is the simplest type of intake structure and is normally used to divert water from small lakes and streams.

*Submerged intakes* are located entirely under water (Figure 41), so they offer no surface obstruction to navigation and are less likely to be damaged by floating debris and ice.

*Movable intakes* (Figure 42) are used in streams that have gently sloping banks and water levels that vary greatly. The intake pump and suction piping are mounted on a dolly that runs on tracks. As the water level rises or falls, sections of discharge piping are removed or added and the rail car is moved accordingly. Movable intakes are also used where good foundation is lacking or where other conditions prevent the construction of a more substantial structure.

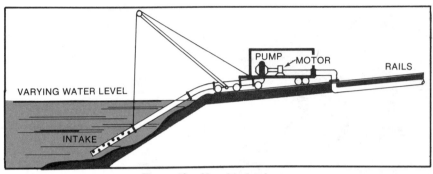

**Figure 42.   Movable Intake**

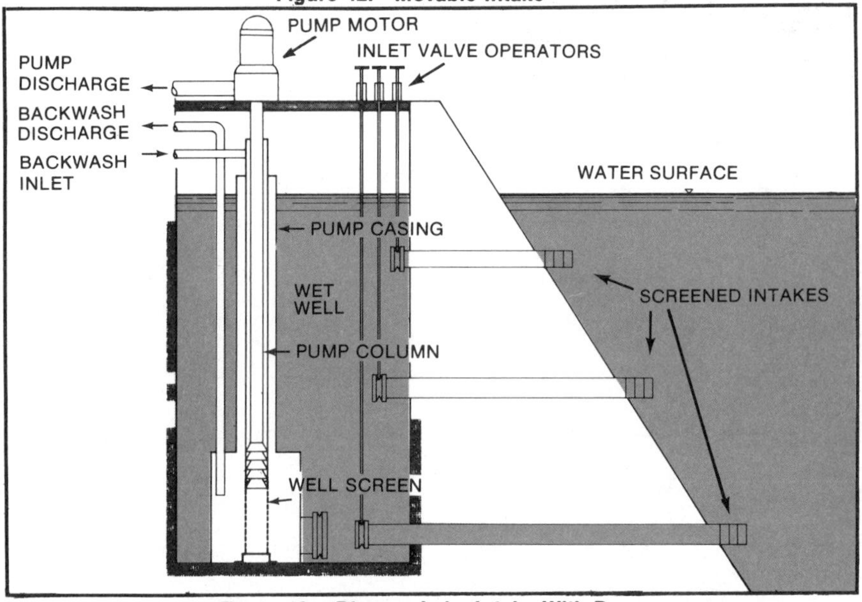

**Figure 43.   River or Lake Intake With Pump**

Reprinted with permission of John Wiley & Sons, Inc., New York, N.Y., from *Elements of Water Supply and Wastewater Disposal*, 2nd ed., by Gordon Fair, John Geyer, and Daniel Okun. Copyright ©1971.

*Pump intake structures* are normally provided with multiple inlet ports (Figure 43). This allows the operator to withdraw water from different depths to compensate for a changing water-surface elevation, to avoid an ice cover, or to select the depth from which to withdraw water of the most favorable temperature or quality.

*Infiltration galleries* are intakes that collect seepage water. They are usually located next to rivers (where they can pick up river water that seeps through the soil) or in the water-bearing materials of river beds. They can also be located in the sides of hills or mountains (where they intercept ground water). In Figure 44, water enters the porous pipe of the gallery by seeping through the river bed deposits. Because it is below ground the infiltration gallery resists freezing. The natural straining action of the river bed screens out debris.

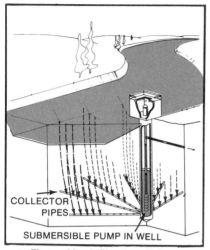

Figure 44. Infiltration Gallery

Figure 45. California Aqueduct

Figure 46. Aqueduct Tunnel

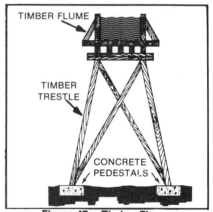

**Figure 47. Timber Flume**
From *Water-Resources Engineering*
by Ray K. Linsley and Joseph B. Franzini.
Copyright ©1964 by McGraw-Hill, Inc.
Used with permission of McGraw-Hill Book Company.

## 4-2. Aqueducts

In the past, the word AQUEDUCT was used to describe any water transmission line. Today the word describes those very large, long, high-volume water transmission lines such as the Catskill Aqueduct, the Colorado River Aqueduct, and the California Aqueduct (Figure 45). The California Aqueduct, for example, is 662 miles long and carries 4 million acre-feet per year—10,000 cubic feet per second, or about 6.5 billion gallons per day. An aqueduct can be an open channel (Figure 45), a tunnel (Figure 46), a surface CONDUIT, a flume (Figure 47), or a combination of these elements. Open aqueducts flow by gravity; covered or closed aqueducts flow either by gravity or pressure. (Some engineers limit the use of the word aqueduct to conduits fabricated entirely in the field.)

## 4-3.  Pipelines and Couplings

PIPELINES are the most common way to transmit water from the raw water source to the treatment plant, and from the plant to the distribution system. A transmission pipeline consists of five major components:

- Pipe
- Pipe couplings
- Valves
- Pumps
- Meters

In the next section you will study types of pipe materials and couplings and in the last three sections valves, pumps, and meters.

Pipe materials commonly used for water transmission today include

- Gray cast iron pipe ................................. CIP
- Ductile iron pipe .................................. DIP
- Steel pipe.......................................... SP
- Asbestos-cement pipe .............................. ACP
- Reinforced concrete pipe .......................... RCP
- Plastic pipe ....................................... PP

You may need to locate a pipeline in the plant or in the field. To do this you must be able to distinguish between various pipe materials. Periodically, sections of pipeline will require replacement and you must know how to handle and couple each type of pipe. You may need to purchase and install new pipelines in the plant to modify plant operation, and this will require that you know which pipe material and coupling method are best suited to your application.

### Cast Iron Pipe—Gray Cast Iron Pipe (CIP) and Ductile Iron Pipe (DIP)

Cast iron pipe has been the standard for underground piping in the US for many years. Two types of cast iron pipe are commonly used by water utilities: gray cast iron pipe (CIP) and ductile iron pipe (DIP). (In this text, "CIP" is used as the abbreviation for *gray* cast iron pipe, not for cast iron pipe in general.)

Gray cast iron is strong but brittle, offers a long service life, and is reasonably maintenance free. The first authenticated installation of gray cast iron pipe was ordered 300 years ago by King Louis XIV of France. (The 15-mile main is still in service today.) There are more than 130 water and gas utilities in the US that have had CIP in service for 100 years or longer. Figure 48 shows a typical CIP transmission line installation.

In contrast to gray cast iron pipe, ductile iron pipe is quite MALLEABLE, combining the corrosion resistance of gray cast iron with the tough mechanical properties similar to steel. DIP has about twice the strength of CIP. Since introduced to the water industry in 1948, DIP has been found especially useful for buried water lines exposed to heavy loads, shocks, and unstable pipe bedding. Because of its strength and toughness DIP is easier to install than CIP. Today over 95 percent of the cast iron pipe manufactured in the US is DIP. Figure 49 gives some indication of the toughness of this pipe material. A similar test would shatter CIP.

*Courtesy of American Cast Iron Pipe Company*
**Figure 48.  Installing Cast Iron Pipe**

*Courtesy of American Cast Iron Pipe Company*
**Figure 49.  Demonstration of Toughness of Ductile Iron Pipe**
A 30-in. diameter ductile iron pipe was severely deformed but did not fail under the impact of a 4500-lb weight dropped from 30 ft.

Because cast iron is metallic, both types of cast iron pipe are subject to attack by corrosive soils. Where corrosive soils are found to be a problem, the outside of cast iron pipe can be economically protected by encasing it in a sleeve of polyethylene plastic or by using standard CATHODIC PROTECTION[18] methods. Coating the pipe with a bituminous tar product also gives protection. The inside surface of unlined cast iron pipe is subject to TUBERCULATION, the pitting and

---

[18] *Basic Science Concepts and Applications,* Chemistry Section, Chemistry of Scaling and Corrosion Control.

*Courtesy of Girard Polly-Pig, Inc.*
**Figure 50.   Television Photograph of Tuberculated Pipe**

growth of NODULES, which reduces the inside diameter and increases the pipe roughness. An example of tuberculated pipe before cleaning is shown in Figure 50. To prevent tuberculation, cast iron pipe is often provided with a cement or a bituminous tar lining. Both linings provide a smoother, longer lasting surface that improves the carrying capacity of the pipe.

Cast iron pipe is easily cut and tapped in the field. It is manufactured in diameters from 3 to 54 in. and in 12, 16, 18, and 20 ft lengths. The pipe is suitable for use with pressures up to 350 psig.[19]

Individual pipe lengths are coupled together with one of the following joints

- Bell and spigot
- Mechanical
- Push-on
- Flanged
- Ball and socket
- Threaded

**Bell and spigot joint.** As shown in Figure 51, the joint is made by slipping the spigot end of the pipe into the bell end. The joint is stuffed with yarn, then sealed with lead as shown in Figure 52. It is not necessary to hold the pipes in perfect alignment when making this joint. A certain amount of misalignment is permissible. As a result this joint can be used when pipe is laid around slight curves. The joint requires skill to construct and does not have a neat appearance. This type of joint is no longer installed in new systems, but you may come across it in old water systems.

**Mechanical joint.** This joint, shown in Figure 53, is made by bolting a follower ring (movable) to the flange on the bell. The follower ring compresses a rubber gasket into place to create a seal. The joint is easily made, requiring no special skill. Since the bell and spigot ends need not fit tightly each joint can be made to deflect a few degrees. Photos of a typical mechanical joint before and after assembly on cast iron pipe are shown in Figure 54.

---

[19]*Basic Science Concepts and Applications,* Hydraulics Section, Pressure.

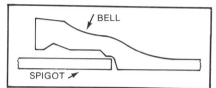

**Figure 51.**
**Partially Completed Bell and Spigot Joints**

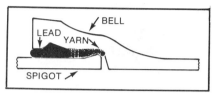

**Figure 52.**
**Completed Bell and Spigot Joint**

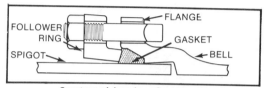

*Courtesy of American Cast Iron Pipe Company*
**Figure 53.   Mechanical Joint**

*Courtesy of Pacific States Cast Iron Pipe Company*
**Figure 54.   Mechanical Joint for Cast Iron Pipe**
A.   Before Assembly    B.   After Assembly

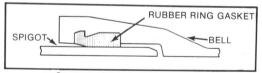

*Courtesy of American Cast Iron Pipe Company*
**Figure 55.   Push-On Joint**

**Push-on joint.** One of the newest joints is the push-on joint for CIP. The joint requires only one part in addition to the specially formed pipe ends—a rubber ring gasket. As shown in Figure 55, this gasket fits into an inside groove within the bell. The spigot end is beveled to avoid damage to the rubber gasket as the pipe is pushed home. The gasket is lubricated to make joining easier and to prevent the rubber ring from rolling out of the groove. The joint is easy to

Courtesy of Cast Iron Pipe Research Association
**Figure 56. Assembling A Push-On Joint**

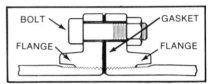

Courtesy of American Cast Iron Pipe Company
**Figure 57. Flanged Joint**

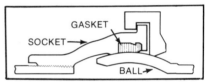

Courtesy of U.S. Pipe
**Figure 58. Flanged Joint in Water Treatment Plant Pipe Gallery**

Courtesy of American Cast Iron Pipe Company
**Figure 59. Ball and Socket Joint**

assemble, watertight, and self-centering. It provides for ample deflection and saves on pipe installation time. Figure 56 shows how a push-on joint is assembled.

**Flanged joint.** The flanged joint shown in Figure 57 has two machined surfaces, which are tightly bolted together with a gasket between them to prevent leakage. This joint is suited for exposed service in water plants, pump stations, and other locations where rigidity, strength, self-restraint, and tightness are required. It is not suited for underground service because unequal settling of the pipe in the trench can cause stresses that may cause the pipe to crack since the joint will not flex. Flanged joints are easy to make and require no special tools. Figure 58 shows a flanged CIP installation in a treatment plant pipe gallery.

**Ball and socket joint.** This is a special purpose joint that provides for a great deal of deflection. The joint, shown in Figure 59, can be used wherever there is a need to abruptly change alignment, such as might occur where a pipeline is routed under a river.

**Threaded joint.** Threaded joints are not often used in CIP and are never used on pipe of over 12-in. diameter. They are costly and must be perfectly aligned before they "catch." Except for the smaller outside diameter of the coupling, threaded joints offer no advantages over flanged joints. In time the threads corrode and the joint is almost impossible to separate. The use of threaded joints is limited to special applications.

## Steel Pipe (SP)

Steel pipe has been in use in water lines in the US since 1852. It is frequently used where pressures are high and large diameter pipe is required. SP is much stronger than CIP, and has about the same or slightly greater strength than DIP; it tends to be lighter than either CIP or DIP. SP resists shock loads and will bend without breaking. It is comparatively inexpensive, easy to install, and more easily transported than CIP; however, SP cannot withstand the external loads that CIP or DIP can. A partial vacuum caused by rapidly emptying SP can cause pipe distortion or complete collapse, as shown in Figure 60.

*Courtesy of City of Los Angeles Dept. of Water & Power*
**Figure 60. Collapsed Steel Pipe**

Because it is metallic, SP is subject to corrosion; and (because of its different chemical composition) corrosion is often more severe in steel than in either DIP or CIP. Corrosion of the thin walls of steel pipe can create high maintenance costs and short pipe life. Special linings and coatings are required to maintain adequate service life.

Bitumastic enamel, a coal-tar material, is widely used to line and coat SP. Cement-mortar lining and coating, epoxy lining and coating, and protective coal-tar wrapping are also accepted methods of corrosion protection. Special care must be taken to prevent damage to these protective coatings. Small scars or breaks in the coatings must be repaired to prevent accelerated corrosion in the area of the break.

There are two methods of manufacturing steel pipe—the mill process and fabrication. Mill pipe is available in diameters from 1/8 to 36 in., whereas fabricated pipe is routinely available in diameters from 4 to 96 in. In special cases steel pipe has been fabricated in diameters of 30 ft or more. Steel pipe lengths are usually 25 to 30 ft.

Steel pipe can be joined by welding. Examples of typical welds are shown in Figure 61. In addition, SP can be joined using

- Push-on joints
- Mechanical joints
- Threaded couplings
- Flanged joints
- Compression or mechanical couplings
- Bell and spigot joints

Examples of most of these joints have been shown earlier in this section.

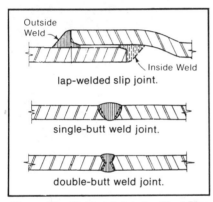

**Figure 61.   Welded Joints for Steel Pipe**

*Courtesy of A/C Pipe Producers Association*

**Figure 62. Handling Asbestos-Cement Pipe**

## Asbestos–Cement Pipe (ACP)

Introduced to North America in 1931, ACP is now widely used in water systems, particularly in those areas where metallic pipe is subject to corrosion.

ACP is made of asbestos fiber and portland cement, combined under pressure to form a dense, strong pipe material. It is highly resistant to corrosion, does not conduct electricity, is relatively light weight (about one quarter the weight of CIP), and is easily cut, handled (as shown in Figure 62), and coupled in the field. ACP is easy to drill and tap, and it is very smooth, offering low resistance to flow. ACP breaks easily if it is flexed in handling or during uneven settlement in the pipe trench. Placing a proper bedding in the trench before placing the pipe (Figure 63) is important in preventing flex. You should be particularly careful when excavating around ACP because it can be easily punctured by excavating tools.

Although ACP is corrosion resistant, the outside can be attacked by acid soil conditions and the interior can be deteriorated by a corrosive water. The pipe

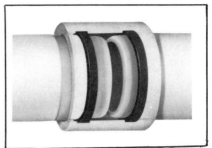

Courtesy of A/C Pipe Producers Association
**Figure 63.  Asbestos-Cement Pipe
Installation With Proper Bedding**

Courtesy of A/C Pipe Producers Association
**Figure 64.  Asbestos-Cement Sleeve
Coupling**

does not conduct electricity, so it is not subject to corrosion by ELECTROLYSIS (natural electric currents moving between the soil and the pipe material). However, because it does not conduct electricity, frozen ACP cannot be thawed electrically.

ACP is manufactured in diameters from 4 to 42 in. with lengths of 10 to 13 ft. Individual pipe lengths are easily connected by sleeve couplings, shown in Figure 64. The rubber rings inside the coupling seat in grooves at the end of each pipe section. This provides a watertight connection yet permits the coupled pipe to deflect as much as 6 deg to accommodate slight irregularities in bedding or to allow the pipe to be installed around slight curves.

## Reinforced Concrete Pipe (RCP)

Reinforced concrete pipe, shown in Figure 65, is a widely used material for water transmission lines, particularly in 48 in. diameter and larger sizes. RCP can be classified into two general types:

- Non-steel cylinder type
- Steel cylinder type

**Non-steel cylinder RCP.** The construction of non-steel cylinder RCP begins by forming one or more cages of circumferential (spiraling around the outside) and longitudinal (lengthwise) reinforcing steel. These cages are then placed in a form and are lined and coated with concrete. The example shown in Figure 66 uses two cages, one near the inside of the pipe and one near the outside. The joint shown in the figure is called a tongue and groove.

Non-steel cylinder RCP is usually manufactured in diameters from 60 to 144 in. and in lengths from 8 to 24 ft. It is designed to be used for very low pressure applications, from 0 to 65 psig.

**Steel Cylinder RCP.** The steel cylinder type of RCP is shown in Figure 67. In a common type of construction, the pipe begins as a plain steel cylinder or tube. Next, steel spigot and steel bell rings are welded to the ends. Then the inside is lined with smooth cement mortar and cement mortar is placed on the outside of the cylinder. Prestressing wire is wrapped around the assembly and cement-mortar coating is added over the wire.

Courtesy of the Concrete Pipe Division of U.S. Pipe and Foundry
**Figure 65.  Installing Concrete Pipe**

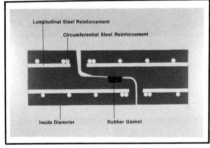

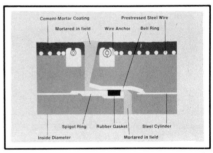

Courtesy of the Concrete Pipe Division of U.S. Pipe and Foundry

**Figure 66.  Non-Steel Cylinder Reinforced Concrete Pipe**

Courtesy of the Concrete Pipe Division of U.S. Pipe and Foundry

**Figure 67.  Steel Cylinder Reinforced Pipe**

This type of steel cylinder RCP is usually manufactured in diameters from 48 to 144 in., and in lengths from 8 to 24 ft. The pipe is suited for low to very high pressure applications, from 40 to 700 psig. Other types of steel cylinder RCP range in diameter from 12 to 180 in. and are available for low to medium pressure applications, up to 400 psig.

RCP is a popular, low maintenance pipe material. In general it is not subject to tuberculation. It can be operated free from corrosion when carrying POTABLE (suitable for drinking) water. However, any concrete product will corrode under acid or alkaline conditions. Concrete does not conduct electricity: This protects it from electrolysis, but also prevents frozen pipe from being electrically thawed.

Because of the compressive strength of concrete, RCP is well suited for installation under high backfill loads. Concrete is a slightly porous material, therefore, non-steel cylinder RCP has a tendency to leak; however, the steel

Courtesy of the Concrete Pipe Division
of U.S. Pipe and Foundry
**Figure 68.  Reinforced Concrete Pipe
Transmission Line Under Construction**

Courtesy of Ethyl Corporation, Pipe Products
Division
**Figure 69.  Handling PVC Pipe**

cylinder version is quite watertight. RCP is more difficult to tap than CIP, and if damaged it is quite difficult to repair. Bell and spigot or push-on joints are used to join RCP. Figure 68 shows an RCP transmission line installation.

## Plastic Pipe (PP)

Plastic is fairly new pipe material; it is slowly gaining acceptance in the water utility field. Used increasingly now in distribution systems, plastic pipe also finds application in small-diameter transmission lines.

Polyvinyl chloride (PVC) is one of the more popular plastic pipes. It is one of the few plastic pipes available in the larger diameters used in transmission lines. Polyvinyl chloride was first used as a pipe material in Germany in the 1930s. Since PVC is non-metallic, it will not corrode from electrolysis or electrochemical action. The chemicals in soils do not damage it. Consequently linings, coatings, cathodic protection, or other corrosion protection are not needed. The pipe cannot be thawed electrically. As shown in Figure 69, PVC is easily carried and handled during joining. The inside surface of PVC is very smooth and offers very little resistance to flow. PVC can easily be cut and assembled without the need for special pipe tools, as shown in Figures 70, 71, 72.

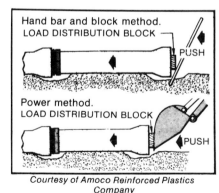

Courtesy of Amoco Reinforced Plastics
Company
**Figure 70.  Assembling PVC Pipe**

Courtesy of Johns-Manville
**Figure 71.  PVC Can Be Cut With Power or
Hand Saw**

1. Clean Bell and End. Be sure no dirt can lodge between the ring and the bell or pipe end.

2. Set Ring in Groove . . . with painted edge facing toward end of bell.

*Courtesy of Johns-Manville*

3. Lubricate Pipe End . . . with a light film of lubricant.

4. Push End in . . . so that reference mark on spigot end is flush with end of bell.

**Figure 72.   Assembling PVC Pipe With Rubber Ring Push-On Joint**

A disadvantage of this pipe material is that the sun's ultraviolet rays will cause it to deteriorate. PVC or any other plastic pipe should always be protected from direct sunlight during storage. If plastic pipe is to be left in an open trench for several days or more, the pipe should be shaded with a small amount of backfill. Plastic pipe can also be damaged by falling rocks during the backfilling operation.

PVC is available in diameters from 3/4 to 12 in., suitable for use under pressures up to 200 psig. The pipe comes in standard lengths of 20 and 40 feet.

There are two commonly used joints for PVC:

- Rubber ring push-on
- Solvent weld

Of the two, the push-on type is most widely used. As shown in Figure 72, the joint is made by inserting a spigot end into a bell end tha* contains a rubber ring. The spigot compresses the rubber ring to create a tight seal.

The solvent weld joint is made using a sleeve-type coupling on plain-end pipe as shown in Figure 73. PVC primer and then PVC cement are applied to the inside of the sleeve and the outside of the plain end. The two pieces must be joined immediately by pushing together and twisting slightly. The solvent cement fuses the pipe to the sleeve creating a rigid, inflexible, watertight seal.

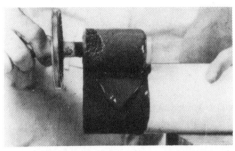

1. Stir the cement and use as is, using the proper size applicator. Apply a *full even layer* of cement on the pipe equal to the depth of socket. Flow the cement on with the applicator—do not brush it out to a thin, paint-type layer that will dry in a few seconds.

2. Apply a *medium layer* of cement to the fitting socket; *avoid puddling* cement in the socket. On bell-end pipe, do not coat beyond the socket depth or allow cement to run down in the pipe beyond the bell.

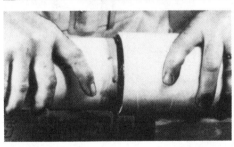

3. Assemble the pipe and fitting without delay. Cement must be wet. Use sufficient force to ensure that the pipe bottoms in the fitting socket. If possible, twist the pipe 1/8 to 1/4 turn as you insert it.

**Figure 73. Solvent Weld PVC Joint**

Reprinted with permission of Industrial Polychemical Service, 17109 S. Main Street, Gardena, CA 90248, from *Solvent Welding PVC* by Industrial Polychemical Service, copyright ©1971.

## 4-4. Valves

The valves discussed in this section fall into two categories. First, there are valves that you will use routinely to

- Stop flow
- Regulate flow (THROTTLE flow)
- Drain the line
- Isolate a section of the line

These valves can be opened or closed by hand or by motorized operators. Motor operated valves may be activated at the valve location or from a distance, through a remote-control circuit.

Second, there are valves used to protect the pipeline. These valves operate automatically to

- Bleed off air
- Take in air
- Eliminate water hammer
- Prevent backflow

Table 12 summarizes the valves that are most suitable for the types of transmission line service noted.

**Table 12.   Valve Application**

| Type of Service | Valves Suitable for This Service | Valve Numbers |
|---|---|---|
| Stop flow | 1, 2, 4, 5 | 1. Gate |
| Regulate flow | 1, 4, 5 | 2. Globe |
| Prevent backflow | 3, 6 | 3. Swing check |
| Line draining | 1, 4, 5 | 4. Butterfly |
| Bleed off air | 7 | 5. Plug |
| Take in air | 7 | 6. Cushioned or restrained check |
| Eliminate water hammer | 8 | 7. Air & vacuum relief |
| Isolate sections of line | 1 | 8. Pressure relief |

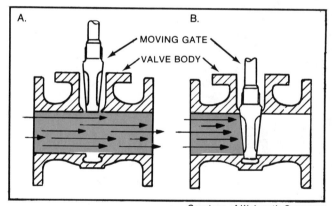

Courtesy of Walworth Company

**Figure 74.   Function of a Gate Valve**
A.   Open Position   B.   Closed Position

Courtesy of DeZurik, a Unit of General Signal
**Figure 75.**
**Typical Gate Valve**

## Gate Valves

A gate valve consists of a circular plate or gate (Figure 74) that is moved down into the valve body by a hand-wheel operator (Figure 75) to effectively stop all flow. This type of valve is widely used in transmission lines to stop flow, to isolate sections of the line, and to drain the line. Gates are rugged, resist leakage, serve well under high pressure and most important, create no HEAD LOSS (do not

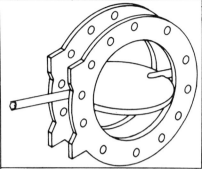

**Figure 76.   Butterfly Valve Detail**

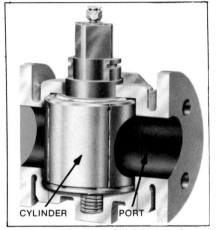

CYLINDER        PORT

*Courtesy of W-K-M Valve Group*
**Figure 78.   Cylinder Type Plug Valve**

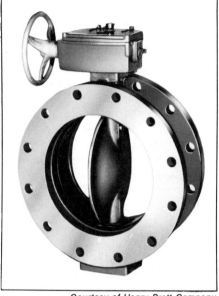

*Courtesy of Henry Pratt Company*
**Figure 77.   Butterfly Valve With Geared Operator**

obstruct flow) when fully open. Because of the tremendous pressures against the gate, valves 16 in. and larger are normally provided with gearing to make opening the valve easier. In some situations the hand-wheel operator is removed and replaced with a motor-driven operator.

## Butterfly Valves

The butterfly valve consists of a movable disc rotating on a spindle and housed in a valve body (Figure 76). When open it offers low head loss. It is quick opening, tight closing, easy to operate and can be used to throttle flows. Under high pressure (above 125 psig) the metal seats do not provide dripless closure—in such cases a rubber or composition seat can be used. Unlike gate valves, butterflies are relatively easy to open under high HEADS (because the pressure pushing on half of the upstream side of the disc tends to force it closed, balancing the pressure on the other half which tends to force it open). Geared operators (Figure 77) or motor-driven operators are often provided on butterfly valves that, due to large size or to location, would be difficult or inconvenient to operate otherwise.

## Plug Valve

A plug valve is a machined-surface cylinder (sometimes a sphere—called a *ball valve*—or a truncated cone) with a bored port or passageway; the cylinder is mounted inside the valve body on a shaft (Figure 78). By turning the shaft 90 deg (a quarter turn) the cylinder will move from open position (shown in Figure 78) to a totally closed position. A wrench-like operating handle is often used to turn the operating nuts of large plug valves. Figure 79 shows a straight-through plug valve using a machined cone instead of a machined cylinder. Figure 80 shows three examples of multiport plug valves. Notice the different types of flow paths that are created by turning these cylinders 90 deg at a time. When installed at crosses or tees, multiport plug valves can be used to divert flow from one pipe branch to another without stopping the flow of water. Like gate valves, round-port plug valves offer no head loss when fully open. Although sediment can prevent gate valves from closing tightly, plug valves are not easily fouled by sediment; as a result, they are better suited for use in pipelines carrying water high in sediment load.

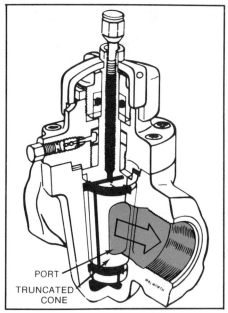

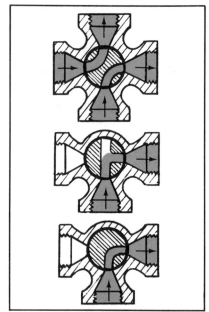

*Courtesy of Walworth Company*

**Figure 79. Truncated-Cone Type Plug Valve**

*Courtesy of Walworth Company*

**Figure 80. Plug Valve Flow Patterns**

## Globe Valves

The globe valve (Figure 81) is commonly used for ordinary household water faucets. It has a circular disc that moves downward into the valve port to shut off flow. Because of the turns the water must make moving through the valve (Figure 82) globe valves produce high head losses when fully open. Therefore, they are not properly suited for transmission line duty, where head loss is critical. Where rapid draining is not important, globes may be used to drain transmission

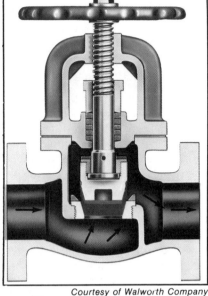

*Courtesy of Walworth Company*
**Figure 81.  Typical Globe Valve**

*Courtesy of Walworth Company*
**Figure 82.  Direction of Flow**

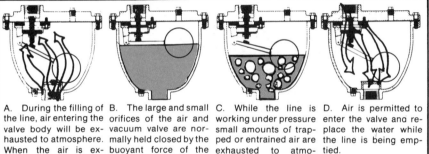

A.  During the filling of the line, air entering the valve body will be exhausted to atmosphere. When the air is expelled, and water enters the valve, the float will rise and cause the orifices to be closed.

B.  The large and small orifices of the air and vacuum valve are normally held closed by the buoyant force of the float.

C.  While the line is working under pressure small amounts of trapped or entrained air are exhausted to atmosphere through the small orifice.

D.  Air is permitted to enter the valve and replace the water while the line is being emptied.

*Courtesy of GA Industries, Inc.*
**Figure 83.  Operation of an Air and Vacuum Relief Valve**

lines. Though quicker to operate and less costly to repair than gates, 3-in. and larger globes are more expensive.

## Air and Vacuum Relief Valves

In long pipelines air can accumulate at the high points in the line. This causes *air binding,* the partial blockage of flow by the entrapped air. Air and vacuum relief valves reduce the problem by automatically venting the unwanted air.

Periodically, pipelines must be drained for routine maintenance or repair. As the water flows out through drain valves a vacuum can be created inside the pipeline. If the vacuum is great enough it can completely collapse the pipeline, as shown previously in Figure 60. The air and vacuum relief valve automatically allows air into the pipe to occupy the volume that was filled by water, so no vacuum is created. This prevents pipe collapse and also shortens draining times.

The diagrams in Figure 83 show a typical air and vacuum relief valve and describe how it works.

## Pressure Relief Valves

WATER HAMMER, the slam or shudder you sometimes hear and feel when a valve is closed quickly, is caused by the rapid increase in pressure as the entire moving column of water within the pipe is suddenly brought to a stop against the valve. Pump start-up or shutdown can also cause water hammer. The increased pressure may be many times the pipeline's normal operating pressure, which can seriously damage valves and cause pipelines to burst or to separate at the joints. Water hammer can be controlled or eliminated by using pressure relief valves (Figure 84). Such automatic valves have a spring tension pre-set to a certain operating pressure. Any greater pressure will open the valve, allowing water to escape and preventing an excess pressure build-up. These valves are suitable for service in small pipelines; in large pipelines the destructive energy of water hammer is absorbed with SURGE TANKS. Even on lines protected with relief valves or surge tanks, sudden valve or pump operation should be avoided.

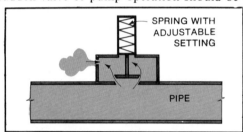

**Figure 84.   Pressure Relief Valve**
From *Water-Resources Engineering* by Ray K. Linsley and
Joseph B. Franzini. Copyright ©1964 by McGraw-Hill, Inc.
Used with permission of McGraw-Hill Book Company.

## Check Valves

Check valves (Figures 85, 86) are automatic valves used to prevent *backflow*. When a pump shuts down, the discharge line (uphill side of the pump) contains water that has just been pumped. Without a check valve this water will drain back through the pump, turning it backward and damaging both pump and motor. Check valves prevent the damage by preventing backflow. They also keep water in the discharge line, providing pressure for the pump to work against when it is restarted. This helps prevent pump motor burnout under no-load conditions. Preventing backflow also saves the time and energy of repumping water that has drained back out of the discharge line.

Allowed to close unrestrained, a check valve will slam shut suddenly, creating water hammer that can damage pipe and valves. A variety of devices are normally installed on check valves to minimize slamming—from simple external restraining springs to automatic slow-closing motorized drives (Figures 87,88).

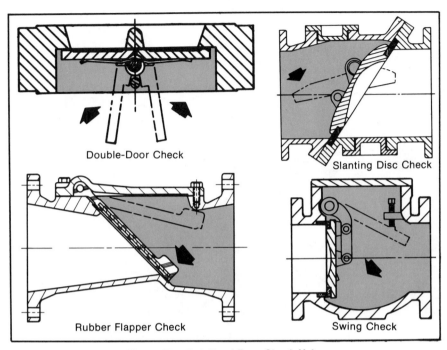

Double-Door Check

Slanting Disc Check

Rubber Flapper Check

Swing Check

**Figure 85.   Details of Four Check Valves**

Reprinted with permission of APCO/Valve & Primer Corp., 1420 So. Wright Blvd., Schaumburg, IL 60193; from *Which Check Valve Should I Use?* by Ralph DiLorenzo, Exec. Vice Pres.; copyrighted ©1975.

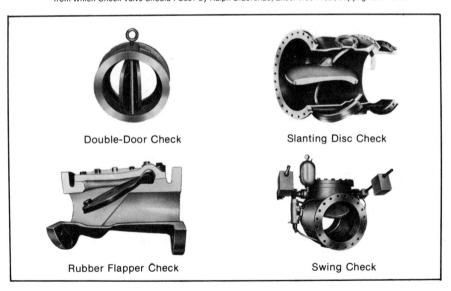

Double-Door Check

Slanting Disc Check

Rubber Flapper Check

Swing Check

**Figure 86.   Photos of Four Check Valves**

Reprinted with permission of APCO/Valve & Primer Corp., 1420 So. Wright Blvd., Schaumburg, IL 60193; from *Which Check Valve Should I Use?* by Ralph DiLorenzo, Exec. Vice Pres.; copyrighted ©1975.

Courtesy of Dresser Manufacturing Division
**Figure 87.  Check Valve
With Spring Restrainer**

Courtesy of Henry Pratt Company
**Figure 88.   Automatic Motor-Operated Check Valve**

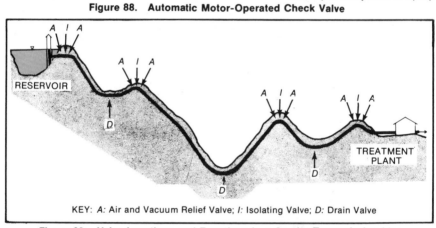

KEY: *A:* Air and Vacuum Relief Valve; *I:* Isolating Valve; *D:* Drain Valve

**Figure 89.   Valve Locations and Functions in a Gravity Transmission Line**

## Typical Valve Installations

Valves are located at set intervals along the transmission line to permit isolating pipeline segments for routine maintenance and repair. Air and vacuum relief valves are located at high points of the system to vent entrapped air and to allow air to enter, preventing a vacuum. Blow-off or line-draining valves are located at low points in the line to permit drainage and sediment removal (see Figure 89). Pressure relief valves are located at low points or other points where excessive pressures could damage the pipeline.

Courtesy of Badger Meter, Inc.
**Figure 91.  Valves and Meter Inside Vault**

Courtesy of Cities Service Company
**Figure 90.  Valve Box With Extended
Operator**

Courtesy of Badger Meter, Inc.
**Figure 92.  Meter Inside Vault**

For small, in-line valve installations, simple valve boxes (Figure 90) provide access to the operating handle of buried valves. For larger installations, particularly those requiring motor operated valves, large concrete vaults house and protect the equipment (Figure 91). Figure 92 shows a typical valve arrangement used for in-line metering installations.

# 4-5.  Pumps

Pumps are a vital part of many water transmission systems, moving water from a lower elevation (from a well, river, or low-level pipeline) to a higher elevation (to a treatment plant, distribution system, or higher-level pipeline). Pump installations vary in size from small, single pumps delivering just a few gallons per minute (Figure 93), to large, multiple-pump installations moving many thousands of gallons per minute (Figure 94).[20]

## Types of Pumps

Pumps can be classified by type of service performed (Table 13) or by principle of operation (Table 14). Although each type of pump in Table 14 has important applications in water systems, only one type, the CENTRIFUGAL PUMP, is commonly used in water transmission systems.

---

[20] *Basic Science Concepts and Applications,* Hydraulics Section, Pumping Problems.

*Courtesy of Worthington Pump, Inc.*
**Figure 93. Small Water Pump Station**

*Courtesy of Worthington Pump, Inc.*
**Figure 94. Large, Multiple-Pump Station**

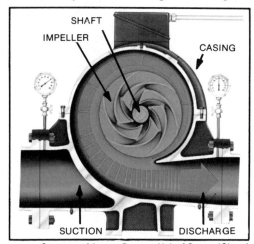

*Courtesy of Aurora Pump, a Unit of General Signal*
**Figure 95. Main Parts of Volute Centrifugal Pump**

## Centrifugal Pumps

As shown in Figure 95, the main parts of a centrifugal pump are

- IMPELLER: a rotating bladed disc that imparts a force to the water being pumped
- SHAFT: a cylindrical rod on which the impeller is mounted. The shaft transmits power from the motor to the impeller
- CASING: the housing surrounding and enclosing the impeller (called a *bowl* in a vertical turbine)
- SUCTION: the opening through which water enters the casing or bowl
- DISCHARGE: the opening through which water leaves the casing or bowl after being pumped

**Table 13.   Pump Classification by Service***

| Service | Function of Service |
|---------|---------------------|
| Low | To lift water from source to treatment processes, or from clear well to filter washing system |
| High | To discharge water under pressure to distribution system |
| Booster | To increase pressure in distribution system or to supply elevated storage tanks |
| Well | To lift water from shallow or deep wells and discharge it to treatment plant or to distribution system |
| Chemical feed | To add chemical solutions at desired dosages to treatment processes |
| Sampling | To pump water from sampling sources to laboratory or automated analyzers on continuous or programmed basis |
| Cooling | To recirculate water through heat exchangers, or to discharge cooling water to sprays or cooling towers |
| Boiler feed | To pump water into steam boilers |
| Fire | To supply extra pressure for fighting fires |

*Copyright 1978 by the Technical Publishing Co. Reprinted by special permission from the manual *Pumps and Pumping, Manual of Practice Number One* of Water and Wastes Engineering.

**Table 14.   Pump Classification by Operating Principle***
**(Alphabetical Listing)**

| Operating Principle | Application (service) |
|---------------------|------------------------|
| Air lift | Deep well pumping |
| Centrifugal | Low service, high service, boiler feed booster, well, sampling, cooling, fire |
| Ejector | Well pumping |
| In-line (centrifugal) | Deep well, booster |
| Peristaltic | Sampling, chemical feed |
| Progressing cavity | Low service, booster, chemical feed, sampling, sludge |
| Reciprocating displacement | Low service, high service, chemical feed, boiler feed, sludge |
| Rotary displacement | Low service, chemical feed, sampling, fire |
| Turbine regenerative | Chemical feed, general purpose |
| Vertical turbine | Well pumping, booster |

*Copyright 1978 by the Technical Publishing Co. Reprinted by special permission from the manual *Pumps and Pumping, Manual of Practice Number One* of Water and Wastes Engineering.

Centrifugal pumps move water by the centrifugal force created when the impeller rotates inside the casing. Based on the internal design of the casing and impeller, centrifugal pumps can be classified into four groups:

- Volute
- Diffuser
- Axial-flow
- Mixed-flow

**Volute centrifugal pumps.** The VOLUTE centrifugal pump, shown in Figure 95, is the most common type of centrifugal used in water systems. It takes its name from the spiral-shaped interior of the casing, called the volute. As the motor (located behind the casing in Figure 95) turns the shaft, the impeller spins. The centrifugal force created by the spinning impeller throws the water outward into the volute. This creates a partial vacuum at the center, or EYE, of the impeller, which pulls more water into the pump from the suction opening.

Pressure builds as more and more water is thrown into the volute area, forcing the water around the spiral and out the discharge. The volute shape changes the high velocity and low pressure head of the water leaving the impeller to a lower velocity and higher pressure head at the discharge. Notice that the movement of water during pumping is radially outward, away from the shaft. Pumps moving water in this direction are called RADIAL-FLOW PUMPS.

The volute centrifugal is a good general service pump. Although it is not easily clogged by the small particles of dirt and organic debris found in raw water, it is best used for pumping clean, clear water.

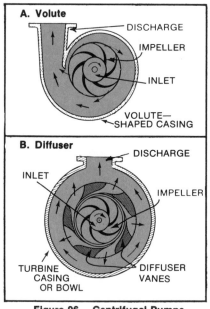

**Figure 96.   Centrifugal Pumps**

Reprinted with permission of Johnson Division, UOP Inc., St. Paul, Minn., from *Ground Water and Wells*, copyright ©1975.

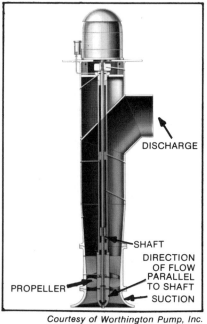

*Courtesy of Worthington Pump, Inc.*

**Figure 97.   Axial-Flow Centrifugal Pump**

**Diffuser centrifugal pumps.** Figure 96 shows the two basic differences between volute and DIFFUSER centrifugal pumps: First, in the diffuser pump the impeller is mounted at the center of a circular casing, so there is no spiral or volute chamber. Second, stationary diffuser vanes are fixed to the inside of the casing. The vanes perform the same function as the volute: they convert the velocity of the outwardly-thrown water to pressure head. Diffuser vanes generally perform the conversion more effectively than the volute; as a result, the diffuser type pump usually operates more efficiently than the volute type. As with volute pumps, the movement of the water during pumping is radially outward, away from the shaft.

The diffuser centrifugal pump is sometimes called a TURBINE pump since regenerative turbine pumps (true turbine pumps) use vanes similar to those in the diffuser casing.

Because of its efficiency the diffuser pump is useful in high-head applications. However, due to the small clearance between the impeller and diffusers, it is easily clogged by dirt. Therefore, its best application is with clean, clear water.

**Axial-flow centrifugal pumps.** The AXIAL-FLOW pump, sometimes called a propeller pump, moves water by the lifting or pushing action of the propeller blades—much the same as the pushing action of an outboard motor propeller. Although commonly referred to as a centrifugal pump, this pump does not rely primarily on centrifugal force to move water. As the name axial implies, water travels through the pump parallel to the shaft, or axis.

The axial-flow pump is not easily clogged and is therefore useful in pumping raw water prior to treatment, particularly in low-head applications. It is not suited for high-head duty. An axial-flow pump is shown in Figure 97.

**Mixed-flow centrifugal pumps.** As explained above, the volute centrifugal throws water radially outward from the shaft, and the axial-flow centrifugal lifts or pushes water parallel to the shaft. In the MIXED-FLOW centrifugal, the pumping action is a combination of radial and axial flows.

Like the axial-flow pump, the mixed-flow centrifugal is not easily clogged and therefore can be used to pump water with some small solids, as might be encountered in raw water intakes.

*Courtesy of Aurora Pump, a Unit of General Signal*
**Figure 98.   Operating Single-Stage Centrifugal Pump**

**Single- and multistage centrifugals.** A centrifugal pump with one impeller and matching casing, like the one shown in Figure 95, is called a SINGLE-STAGE PUMP. A typical operating single-stage centrifugal pump is shown in Figure 98. A MULTISTAGE PUMP has two or more impellers and casings arranged so the discharge from one impeller enters the eye of the next impeller. A single-stage centrifugal pump is commonly used to create heads up to 250 ft. Higher heads can be achieved by adding stages (impellers and casings).

**Vertical turbine pumps.** In VERTICAL TURBINE centrifugal pumps the impellers are mounted on a vertical shaft. These pumps take water in at the bottom of the pump assembly and discharge it near the top. Vertical turbines may be designed as radial, axial, or mixed flow centrifugals. The most common vertical turbine pumps are a type of diffuser-centrifugal like the one shown in Figure 96B. These pumps are usually multistage. Figure 100 shows a turbine impeller.

The main parts of the pump, as listed at the beginning of the pump section, are shown in Figure 99; also shown are the BOWL DISCHARGE, DISCHARGE COLUMN, LINE SHAFT, and drive motor.

The drive motor turns the line shaft, which turns the impellers inside the bowls. The first (bottom) impeller draws water up the suction pipe and into the impeller eye (Figure 100). The centrifugal force created by the spinning impeller

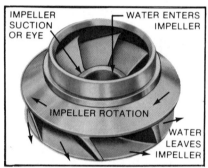

Courtesy of Worthington Pump, Inc.
**Figure 100.   Turbine Impeller**

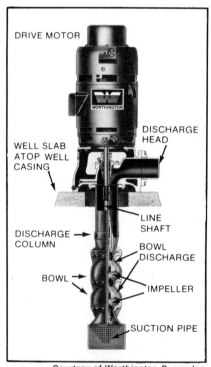

Courtesy of Worthington Pump, Inc.
**Figure 99.   Two-Stage Short-Coupled
Turbine Pump**

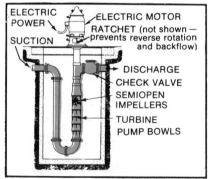

**Figure  101.   Turbine  Booster  Pump
Installation**

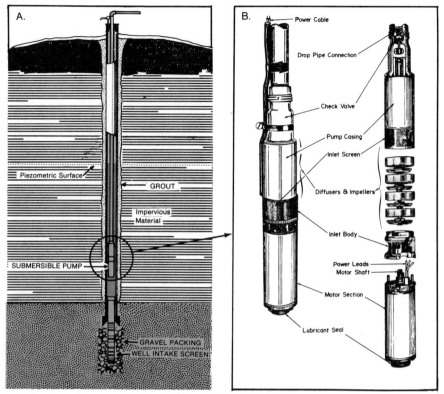

**Figure 102. Submersible Pump**
A. Installed in Well    B. Main Parts

throws the water outward against the bowl, which has a diameter only slightly greater than the impeller. Pathways, cast into the bowl wall, act as diffusers (changing velocity to pressure) while channeling the flow upward into the eye of the next impeller. This process is repeated until the water leaves the last, or top, bowl and enters the discharge column. The water then travels up through the discharge column to the surface, where it leaves through the discharge. In addition to serving as the pipeline for pumped water, the discharge column supports the weight of the submerged pump assembly and houses the line shaft.

Multistage vertical turbine pumps have a long, compact shape that is ideal where space is limited and high lifts are involved. They are used widely as water supply intake pumps from lakes, rivers, and reservoirs, and in deep-well installations. In such applications the raw water should be screened to remove even the smallest of particles. This is because vertical turbines are easily clogged or damaged due to the small clearance between the impeller and the bowl. Vertical turbines also find application as booster pumps, installed along transmission lines and in distribution systems. The pump shown in Figure 99, for example, can be used as a booster pump. A typical booster pump installation is shown in Figure 101.

Because of their high lifting capacity, multistage vertical turbine pumps are often used in transmitting water from the well to the point of treatment. When used this way the well pumps are an integral part of the transmission system. Even though vertical turbines are compact, installation of any pump in a well requires that the well have a larger diameter than would be needed for a simple intake pipe connected to an above-ground pump. When vertical turbine pumps are installed in wells, the drive motor, usually electric, may be installed directly on top of the well casing, as shown in Figure 99. In some installations, however, the motor is located in the well itself, just below the pump assembly. An example of this type of pump, called a submersible pump, is shown in Figure 102.

**Centrifugal pump advantages and disadvantages.** Centrifugal pump advantages and disadvantages include:

*Advantages*
- Low levels of operating noise and vibration
- Simplicity of operation
- Low purchase price
- Ease of mounting—can be connected directly to the drive motor
- Ease and low cost of maintenance and repair
- Stability—can produce a steady, uniform discharge
- Versatility—available for either horizontal or vertical mounting
- High efficiency (when operated at *design point*[21])

*Disadvantages*
- Limited suction lift (distance from impeller eye down to level of water being pumped cannot exceed about 12 ft)
- Lack of self-PRIMING capability (when started dry, pump will not pull water up into itself without special equipment
- Susceptibility to air leaks on the suction side—these reduce performance

**Centrifugal pump operating tips.** For long service life and good performance, certain procedures must be followed when starting, running, and stopping centrifugal pumps.

*Starting.* Centrifugal pumps should always be started with the discharge valve *closed* or at least partially closed (throttled)—even in an installation with a check valve. This will protect the motor from burn-out (see *Stopping, item 3, below*) and reduce the power demand at start-up. Open the discharge valve slowly after the pump is running.

*Running.* The abrasive action of sand will greatly shorten the life of any pump, and dirt or sand may clog turbine pumps. If a well is producing sand, the well screen may need to be replaced. This should be done immediately or else the pump will be severely damaged.

Where lubricated pump parts come in contact with a drinking water supply,

---

[21] *Basic Science Concepts and Applications,* Hydraulics Section, Pumping Problems (Reading Pump Curves).

the lubricant used should be water or a non-toxic substance such as mineral oil. *Stopping.* When a centrifugal pump is turned off, the water in the discharge line can flow back through the pump, turning the pump backward. This presents three problems: (1) turning backward can damage a pump and motor; (2) when the pump is started again, energy will be wasted repumping water to refill the discharge line; (3) when a centrifugal pump is allowed to pump water into an empty line, the lack of back pressure on the pump may cause the motor to run too fast, overheat, and perhaps burn out. The solution to all these problems is to make sure that a check valve is located in the discharge line near the pump and that it is in proper operating condition. The vertical turbine booster pump installation in Figure 101 shows the necessary check valve that provides backflow protection; also shown is the location of a non-reversing ratchet that gives additional protection by preventing the pump and motor from turning backward.

## 4-6.  Flow Measurement

There is no single measurement an operator makes that is more necessary or useful than flow rate. It is essential that you know exactly the rate at which water is being delivered to the treatment plant or to the community distribution system. You will use this information to determine the correct amounts of treatment chemicals needed, to measure the loading on various treatment units, to determine the adequacy of your raw water supply, to report to local and state agencies, to comply with water codes that require periodic reporting of water diversion, and to make sure you supply the amount of water needed by your community.

Flow measurement devices (flow meters) in the water industry can be divided into two groups based on the volume of flow:

● Meters used to measure large flow volumes, such as the amount of water supplied to a community

● Meters used to measure small flow volumes, such as chlorine feed rates, chemical feed rates, and flows to individual households

This section focuses on large-volume flow meters, suitable for use in water transmission. Volume 2 will cover a broader range of flow meters, both large and small volume types, suitable for use in treatment plants. Volume 3 will cover those types used to measure flow rates to individual households.

There are two types of flow meters used in water transmission: (1) those that measure flow in pressure pipe, and (2) those that measure flow in open channels, flumes, and streams. The more common types are

*Pressure Pipe*
● Venturi meters
● Magnetic meters
● Turbine meters
● Propeller meters

*Open Channel*
● Weirs
● Metering flumes

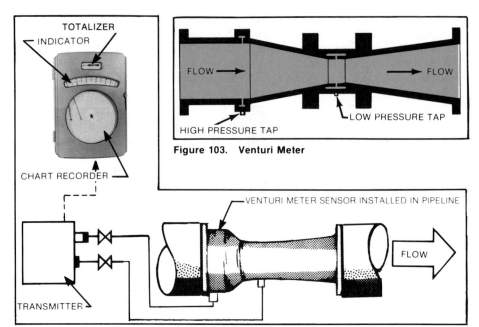

Figure 103.   Venturi Meter

**Figure 104.   Automated Flow Metering System Using Venturi Meter Sensor**   Courtesy of BIF

## Venturi Meters

VENTURI METERS, along with FLOW NOZZLES and ORIFICE-METERS, fall into the category of PRESSURE DIFFERENTIAL METERS, which measure flow by creating and measuring a difference in pressure at two points in the meter. The difference in pressure is directly related to flow rate.

A venturi meter is a totally enclosed special section of pipe for measuring flow rates in pipelines carrying water under pressure. The pipe is shaped like an hourglass to create a throat and is equipped with pressure taps for manual or automatic sensing of pressure at two points (Figure 103).

The amount of water passing through a venturi meter is determined by comparing the low pressure at the throat with the high pressure upstream of the throat. The difference in pressure can be converted to flow rate in cfs or mgd.[22]

Although converting pressure differential to flow can be done with tables or equations, it is most often done automatically and continuously. Instruments sense the pressure differential and electronically convert the differential to a flow signal. The flow signal operates a recorder that automatically compiles a continuous record of flow rates. Figure 104 shows an example of an automated flow meter system.

The major advantage of the venturi is that with minimum maintenance it performs reliably and is reasonably trouble free. The meter does create a head loss, but the loss is relatively small compared with other pressure differential meters.

---

[22]*Basic Science Concepts and Application,* Hydraulics Section, Flow Rate Problems.

## Magnetic Flow Meters

The MAGNETIC FLOW METER is a relatively new type of flow measuring device, developed since 1955. Follow Figure 105 as the principle of the magnetic flow meter is explained.

The magnetic meter looks like a short section of flanged pipe. The inside of the pipe has a smooth insulating liner with no protrusions into the flow stream. Between the liner and the pipe are two magnetic coils. When an electric current is passed through the coils it creates an electromagnetic field around the pipe. The water moving through this magnetic field creates, or *induces,* a small electric current that increases in proportion to the increases in water flow. This small electric current is measured and converted electronically to a measurement of water flow rate.

Magnetic meters are readily available in pipe sizes from 2 to 24 in. and in larger sizes on special order. One of the main advantages of the magnetic meter is that no head loss is created because the meter offers no obstruction to flow. The disadvantage is high cost. Figure 106 shows a photograph of a magnetic flow meter.

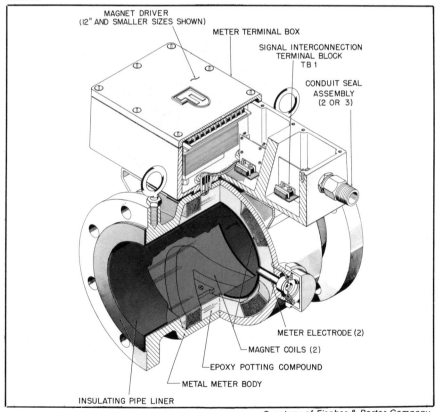

*Courtesy of Fischer & Porter Company*

**Figure 105.   Magnetic Flow Meter**

Courtesy of Fischer & Porter Co.
**Figure 106. Magnetic Flow Meter**

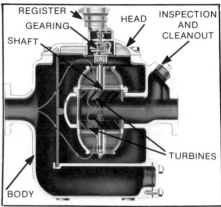

Courtesy of Badger Meter, Inc.
**Figure 107. Turbine Flow Meter**

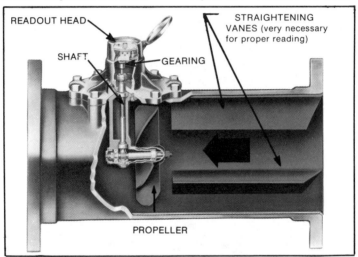

**Figure 108. Propeller Flow Meter**

## Turbine Meters

TURBINE METERS (Figure 107) for large flows are usually BYPASS (or PROPOR-TIONAL) METERS, in which a small portion of the flow in the main pipeline is diverted through a bypass chamber. The diverted flow, which varies in proportion to variations in the main flow, spins a turbine wheel. The spinning wheel generates an electric current, which also varies with the main flow, and which operates a recorder reading directly in flow rate. Turbine meters are quite accurate, and the bypass type creates little head loss. Bypass meters are, however, difficult to maintain.

## Propeller Meters

In PROPELLER METERS (Figure 108) a propeller instead of a turbine is spun by the water. The propeller may be mounted in a bypass chamber or in the main line. Head loss with a bypass propeller meter may be less than with a turbine, but propeller meters are usually less accurate than turbines.

## Weirs

A WEIR is an obstruction to flow placed in an open channel, such as a river, stream, or canal. The entire flow of the channel passes through a carefully sized and shaped notch in the weir. The depth to which the water rises as it builds up behind (upstream of) the weir is directly proportional to the flow.

Unlike pressure pipe meters, which must be maintained and calibrated by specialists, weirs are usually maintained and calibrated—and sometimes constructed—by operators. Therefore it will be helpful for you to understand the principles of measuring flow with weirs.

As shown in Figure 109 there are two types of metering weirs—RECTANGULAR WEIRS and V-NOTCH WEIRS. Within these two general types there are a variety of modifications.

**Rectangular weirs.** Rectangular weirs come in a variety of configurations. The *contracted* rectangular weir shown in Figure 109 is so named because the width of the rectangular notch $L$ is less than the width of the channel. If the notch were as wide as the channel (in which case it would appear as if there were no notch at all) the weir would be called a *suppressed* rectangular weir. Rectangular weirs can be either broad-crested or sharp-crested. A BROAD-CRESTED WEIR (Figure 110) is a weir with a CREST width $b$ somewhere in the range of 6 in. to 15 ft.

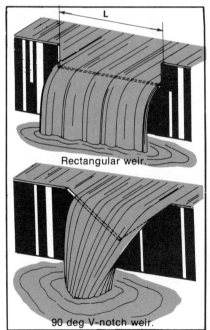

Rectangular weir.

90 deg V-notch weir.

*Adapted from* Manual of Instruction for Water Treatment Plant Operators, *New York State Dept. of Health*

**Figure 109.    Two Types of Weirs**

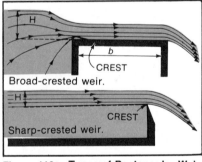

Broad-crested weir.

Sharp-crested weir.

**Figure 110.    Types of Rectangular Weir Crests**

*Courtesy of Leupold and Stevens, Inc.*

**Figure 111.  Sharp-Crested Cippoletti Weir Installation**

Broad-crested weirs are often made of concrete as part of a metering structure built across small creeks or streams.

The more common type of weir is the SHARP-CRESTED WEIR. It can be made of fiberglass, corrosion-resistant metal plate, or wood. A typical installation is shown in Figure 111. The weir plate shown is called a Cippoletti weir; it is a slight modification of the contracted rectangular weir.

**V-Notch Weirs.** V-notch weirs can be made in a variety of different notch angles including 30, 45, 60, and 90 deg. Large angle notches meter higher volume flows than the smaller angle notches. The example in Figure 109 is a 90 deg **V**-notch weir. **V**-notch weirs are always sharp-crested and are used to measure much smaller flows than rectangular weirs since for large flows the size of the **V** becomes unmanageable. To convert a rectangular weir installation like the one shown in Figure 111 to a smaller capacity **V**-notch weir, you need only unbolt the Cippoletti weir plate and replace it with a **V**-notch plate (Figure 112).

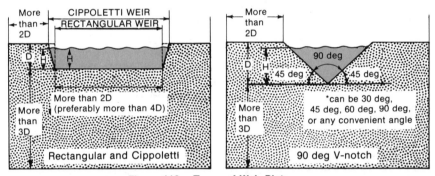

**Figure 112.   Types of Weir Plates**

**Measuring flow with a weir.** The rate of flow passing over the crest of any weir is directly related to the depth of water measured from the weir crest to the water surface. This depth is shown as $H$ in Figure 110. The depth measurement is always made some distance upstream of the weir, usually a distance of $3H$ or more, so that the measurement is not affected by the sloping surface of the water approaching the weir as shown in Figure 110.

The measurement can be read manually from a calibrated rod or STAFF GAGE mounted upstream of the weir (Figure 113). Or the measurement can be made and recorded automatically using a STILLING WELL, float, and recorder (Figure 114). As shown in Figure 115, the float moves up and down as flow increases and decreases. The float turns a pulley, which in turn moves a pen across a moving chart recorder. The system is calibrated to record rate of flow. Figure 116 shows a float-type flow meter and identifies where the flow is totalized (added), indicated, and recorded.

Automatic recording is the best means of monitoring flow rate over a weir. However, once the gage depth $H$ has been measured, the flow rate can be calculated using the appropriate weir flow equation, or it can be read from weir tables.[23] This procedure is often used to check the calibration of the automatic system.

---

[23] *Basic Science Concepts and Applications,* Hydraulics Section, Flow Rate Problems.

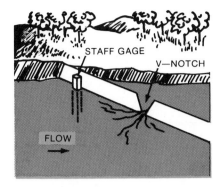

**Figure 113.   Staff Gage Measuring Weir Flow**

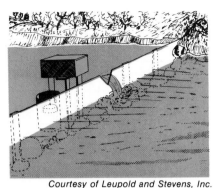

*Courtesy of Leupold and Stevens, Inc.*

**Figure 114.   Stilling Well and Recorder for Measuring Gage Depth H**

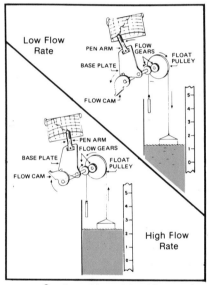

*Courtesy of Leupold and Stevens. Inc*

**Figure 115.   Function of a Float Recorder**

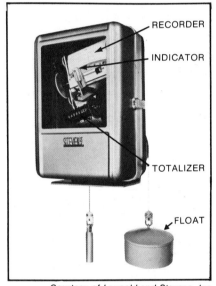

*Courtesy of Leupold and Stevens, Inc*

**Figure 116.   Float-type Flow Recorder**

Assume the staff gage in Figure 113 is measuring the depth of flow over a 90-deg **V**-notch weir. The depth reading, if you look carefully, is 0.22 ft. Using Table 15 you can determine the flow rate. Enter the table under *Head* at *0.2 ft.* Move to the right until you reach the column marked *0.02* (0.2 + 0.02 = 0.22). Now read the flow as *0.057 cfs (0.037 mgd)*.

## Flumes

A FLUME is a specially designed open-channel section (Figure 117). Like a weir, a METERING FLUME can be maintained and calibrated by operators, so it will be helpful for you to understand its principles. The flume offers several advantages over weirs: there are no vertical walls or obstructions across the flow stream to

Table 15.   Discharge of 90-deg V-Notch Weirs*

| Head | .00 | | .01 | | .02 | | .03 | | .04 | | .05 | | .06 | | .07 | | .08 | | .09 | |
|---|---|---|---|---|---|---|---|---|---|---|---|---|---|---|---|---|---|---|---|---|
| Ft. | CFS | MGD | CFS | MGD | CFS | MGD | CFS | MGD | CFS | MGD | CFS | MGD | CFS | MGD | CFS | MGD | CFS | MGD | CFS | MGD |
| 0.1 | .008 | .005 | .010 | .006 | .012 | .008 | .015 | .010 | .018 | .012 | .022 | .014 | .026 | .017 | .030 | .019 | .034 | .022 | .039 | .025 |
| 0.2 | .045 | .029 | .051 | .033 | .057 | .037 | .063 | .041 | .071 | .046 | .078 | .050 | .086 | .056 | .095 | .061 | .104 | .067 | .113 | .073 |
| 0.3 | .123 | .080 | .134 | .086 | .145 | .094 | .156 | .101 | .169 | .109 | .181 | .117 | .194 | .126 | .208 | .135 | .223 | .144 | .237 | .153 |
| 0.4 | .253 | .163 | .269 | .174 | .286 | .185 | .303 | .196 | .321 | .207 | .340 | .219 | .359 | .232 | .379 | .245 | .399 | .258 | .420 | .272 |
| 0.5 | .442 | .286 | .464 | .300 | .487 | .315 | .511 | .330 | .536 | .346 | .561 | .362 | .587 | .379 | .613 | .396 | .640 | .414 | .668 | .432 |
| 0.6 | .697 | .451 | .727 | .470 | .757 | .489 | .788 | .509 | .819 | .529 | .852 | .550 | .885 | .579 | .919 | .594 | .953 | .616 | .989 | .639 |
| 0.7 | 1.02 | .662 | 1.06 | .686 | 1.10 | .711 | 1.14 | .736 | 1.18 | .761 | 1.22 | .787 | 1.26 | .814 | 1.30 | .841 | 1.34 | .868 | 1.39 | .896 |
| 0.8 | 1.43 | .925 | 1.48 | .954 | 1.52 | .984 | 1.57 | 1.01 | 1.62 | 1.04 | 1.66 | 1.08 | 1.71 | 1.11 | 1.76 | 1.14 | 1.82 | 1.17 | 1.87 | 1.21 |
| 0.9 | 1.92 | 1.24 | 1.97 | 1.28 | 2.03 | 1.31 | 2.08 | 1.35 | 2.14 | 1.38 | 2.20 | 1.42 | 2.26 | 1.46 | 2.32 | 1.50 | 2.38 | 1.54 | 2.44 | 1.58 |
| 1.0 | 2.50 | 1.62 | 2.56 | 1.66 | 2.63 | 1.70 | 2.69 | 1.74 | 2.76 | 1.78 | 2.82 | 1.82 | 2.89 | 1.87 | 2.96 | 1.91 | 3.03 | 1.96 | 3.10 | 2.00 |
| 1.1 | 3.17 | 2.05 | 3.24 | 2.10 | 3.32 | 2.14 | 3.39 | 2.19 | 3.47 | 2.24 | 3.55 | 2.29 | 3.62 | 2.34 | 3.70 | 2.39 | 3.78 | 2.44 | 3.86 | 2.50 |
| 1.2 | 3.94 | 2.55 | 4.03 | 2.60 | 4.11 | 2.66 | 4.19 | 2.71 | 4.28 | 2.77 | 4.37 | 2.82 | 4.45 | 2.88 | 4.54 | 2.94 | 4.63 | 2.99 | 4.72 | 3.05 |
| 1.3 | 4.82 | 3.11 | 4.91 | 3.17 | 5.00 | 3.23 | 5.10 | 3.30 | 5.20 | 3.36 | 5.29 | 3.42 | 5.39 | 3.48 | 5.49 | 3.55 | 5.59 | 3.61 | 5.69 | 3.68 |
| 1.4 | 5.80 | 3.75 | 5.90 | 3.81 | 6.01 | 3.88 | 6.11 | 3.95 | 6.22 | 4.02 | 6.33 | 4.09 | 6.44 | 4.16 | 6.55 | 4.23 | 6.66 | 4.30 | 6.77 | 4.38 |
| 1.5 | 6.89 | 4.45 | 7.00 | 4.53 | 7.12 | 4.60 | 7.24 | 4.68 | 7.36 | 4.75 | 7.48 | 4.83 | 7.60 | 4.91 | 7.72 | 4.99 | 7.84 | 5.07 | 7.97 | 5.15 |
| 1.6 | 8.09 | 5.23 | 8.22 | 5.31 | 8.35 | 5.40 | 8.48 | 5.48 | 8.61 | 5.56 | 8.74 | 5.65 | 8.88 | 5.74 | 9.01 | 5.82 | 9.15 | 5.91 | 9.28 | 6.00 |
| 1.7 | 9.42 | 6.09 | 9.56 | 6.18 | 9.70 | 6.27 | 9.84 | 6.36 | 9.98 | 6.45 | 10.1 | 6.55 | 10.3 | 6.64 | 10.4 | 6.73 | 10.6 | 6.83 | 10.7 | 6.93 |
| 1.8 | 10.9 | 7.02 | 11.0 | 7.12 | 11.2 | 7.22 | 11.3 | 7.32 | 11.5 | 7.42 | 11.6 | 7.52 | 11.8 | 7.62 | 11.9 | 7.73 | 12.1 | 7.83 | 12.3 | 7.93 |
| 1.9 | 12.4 | 8.04 | 12.6 | 8.15 | 12.7 | 8.25 | 12.9 | 8.36 | 13.1 | 8.47 | 13.3 | 8.58 | 13.4 | 8.69 | 13.6 | 8.80 | 13.8 | 8.91 | 14.0 | 9.03 |
| 2.0 | 14.1 | 9.14 | 14.3 | 9.25 | 14.5 | 9.37 | 14.7 | 9.49 | 14.9 | 9.60 | 15.0 | 9.72 | 15.2 | 9.84 | 15.4 | 9.96 | 15.6 | 10.1 | 15.8 | 10.2 |

*Source: Stevens Water Resources Data Book, 1st Edition. Formula: CFS = $2.50\ H^{5/2}$    MGD = CFS x .646

*Courtesy of BIF*
**Inset: Fiberglass Parshall Flume Liner**

**Figure 117.    Parshall Flume in Use**

collect silt or debris. The velocity through the flume is high, which helps to keep the flume clean. A disadvantage is that flume installations are more expensive than weirs. The most commonly used flume is the PARSHALL FLUME.

The capacity of the flume is determined by its throat width $W$ (Figure 118). Flumes are available or can be built with throat widths from 1 in. (measuring flows from 0.01 cfs to 0.2 cfs) to 50 ft (measuring flows from 25 cfs to 3000 cfs).

**Measuring flow with a Parshall flume.** The depth of flow at a particular point in the Parshall flume is directly related to the flow rate through the flume. The point at which depth measurements are made is always 2/3 the $A$ distance upstream from the throat (Figure 118).

As with weirs, flume depth measurements can be made using a staff gage attached at the 2/3 $A$ point, or they can be made automatically. If done

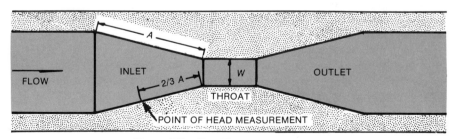

**Figure 118.    Top View of Parshall Flume**

**Figure 119.   Parshall Flume With Staff Gage and Automatic Recorder**

automatically, the depth measurement device usually sends a signal to a flow recorder. The recorder converts the signal to a flow rate measurement and continuously records flow rate in either cubic feet per second or million gallons per day. Figure 119 shows an installed Parshall flume, complete with a staff gage for manual depth measurements and an automatic depth gage and flow recorder.

When depth measurements are made with a staff gage, tables are available to convert readings to flow in cfs or mgd.[24] Table 16 is an example. Assume that you have a Parshall flume with a 9-in. throat width and a staff-gage water-depth reading of 0.61 ft. To determine the flow rate through the flume, enter Table 16 under *Head* at *0.61 ft.* Move across the table to the throat-width column titles *9 in.* Then read the flow rate as *1.44 cfs (0.93 mgd).*

---

[24] *Basic Science Concepts and Applications,* Hydraulics Section, Flow Rate Problems.

Table 16. Flow Rate Through Parshall Flumes*

| Head | DISCHARGE THROUGH THROAT WIDTH, W, OF — | | | | | | | | | | | | |
|---|---|---|---|---|---|---|---|---|---|---|---|---|---|
| | 1 in. | | 2 in. | | 3 in. | | 6 in. | | 9 in. | | 12 in. | | 18 in. | |
| Ft. | CFS | MGD | CFS | MGD | CFS | MGD | CFS | MGD | CFS | MGD | CFS | MGD | CFS | MGD |
| .46 | .101 | .065 | .203 | .131 | .299 | .193 | .61 | .39 | .94 | .61 | 1.23 | .79 | 1.82 | 1.18 |
| .47 | .105 | .068 | .210 | .136 | .309 | .200 | .63 | .41 | .97 | .63 | 1.27 | .82 | 1.88 | 1.22 |
| .48 | .108 | .070 | .217 | .140 | .319 | .206 | .65 | .42 | 1.00 | .65 | 1.31 | .85 | 1.94 | 1.25 |
| .49 | .112 | .072 | .224 | .145 | .329 | .213 | .67 | .43 | 1.03 | .67 | 1.35 | .87 | 2.00 | 1.29 |
| .50 | .115 | .074 | .230 | .149 | .339 | .219 | .69 | .45 | 1.06 | .69 | 1.39 | .90 | 2.06 | 1.33 |
| .51 | .119 | .077 | .238 | .154 | .350 | .226 | .71 | .46 | 1.10 | .71 | 1.44 | .93 | 2.13 | 1.38 |
| .52 | .123 | .079 | .245 | .158 | .361 | .233 | .73 | .47 | 1.13 | .73 | 1.48 | .96 | 2.19 | 1.42 |
| .53 | .126 | .081 | .253 | .164 | .371 | .240 | .76 | .49 | 1.16 | .75 | 1.52 | .98 | 2.25 | 1.45 |
| .54 | .130 | .084 | .260 | .168 | .382 | .247 | .78 | .50 | 1.20 | .78 | 1.57 | 1.01 | 2.32 | 1.50 |
| .55 | .134 | .087 | .268 | .173 | .393 | .254 | .80 | .52 | 1.23 | .79 | 1.62 | 1.05 | 2.39 | 1.54 |
| .56 | .138 | .089 | .275 | .178 | .404 | .261 | .82 | .53 | 1.26 | .81 | 1.66 | 1.07 | 2.45 | 1.58 |
| .57 | .141 | .091 | .283 | .183 | .415 | .268 | .85 | .55 | 1.30 | .84 | 1.70 | 1.10 | 2.52 | 1.63 |
| .58 | .145 | .094 | .290 | .187 | .427 | .276 | .87 | .56 | 1.33 | .86 | 1.75 | 1.13 | 2.59 | 1.67 |
| .59 | .149 | .096 | .298 | .193 | .438 | .283 | .89 | .58 | 1.37 | .89 | 1.80 | 1.16 | 2.66 | 1.72 |
| .60 | .153 | .099 | .306 | .198 | .450 | .291 | .92 | .59 | 1.40 | .90 | 1.84 | 1.19 | 2.73 | 1.76 |
| .61 | .157 | .101 | .314 | .203 | .462 | .299 | .94 | .61 | 1.44 | .93 | 1.88 | 1.22 | 2.80 | 1.81 |
| .62 | .161 | .104 | .322 | .208 | .474 | .306 | .97 | .63 | 1.48 | .96 | 1.93 | 1.25 | 2.87 | 1.85 |
| .63 | .165 | .107 | .330 | .213 | .485 | .313 | .99 | .64 | 1.51 | .98 | 1.98 | 1.28 | 2.95 | 1.91 |
| .64 | .169 | .109 | .338 | .218 | .497 | .321 | 1.02 | .66 | 1.55 | 1.00 | 2.03 | 1.31 | 3.02 | 1.95 |
| .65 | .173 | .112 | .347 | .224 | .509 | .329 | 1.04 | .67 | 1.59 | 1.03 | 2.08 | 1.34 | 3.09 | 2.00 |

*Source: Stevens Water Resources Data Book, 1st edition.

## Selected Supplementary Readings

### Intake Structures

Babbitt, Harold E.; Doland, James J.; & Cleasby, John L. *Water Supply Engineering.* McGraw-Hill Book Company (1962). pp. 123-131.
*Johnston Vertical Pump Application Manual.* Johnston Pump Company, Glendora, CA (1977).

### Aqueducts

Clark, John W. & Viessman, Warren J. *Water Supply and Pollution Control.* International Textbook Co. (1965). pp. 104-110.

### Pipelines and Fittings

*Handbook of Cast Iron Pipe.* Cast Iron Pipe Research Association, 3rd Ed. (1967). Sec. 1-4.
*Steel Pipe Design and Installation.* AWWA (1964). pp. 1-14.
*Design Manual, Johns-Manville Transite Transmission Pipe 18 in.-36 in.* Johns-Manville, Ken Caryl Ranch, Denver, CO 80217. pp. 13-16, 31, 49-54.
*Handbook of PVC Pipe, Design and Construction.* Uni-Bell Plastic Pipe Association, 2655 Villa Creek Drive, Suite 164, Dallas, TX 75234. pp. 2-5, 18-22.
*Installation of Concrete Pipe.* AWWA (1961). pp. 1-7.
*PVC Pipe Installation Guide.* Amoco Chemicals Corporation, 1530 Commerce Drive, Stow, OH 44224.
*Techite Pipe Installation Guide.* Amoco Reinforced Plastics Company, 3100 Jefferson St., Riverside, CA.

### Valves

Babbitt, Harold E.; Doland, James J.; & Cleasby, John L. *Water Supply Engineering.* McGraw-Hill Book Company (1962). pp. 248-273.
*Water Distribution Operator Training Handbook.* AWWA (1976). pp. IX-1 to IX-13.
Lyons, Jerry L. & Askland, Carl L., Jr. *Lyon's Encyclopedia of Valves.* Van Nostrand Reinhold Company (1975). pp. 3-84.

### Pumps

Walker, Rodger. *Pump Selection.* Ann Arbor Science Publishers, Inc. Ann Arbor MI. (1972). pp. 3-22.
*Johnston Vertical Pump Application Manual.* Johnston Pump Company, Glendora, CA (1977).
*Technical Data.* Worthington Pump Corporation, 14 Fourth Ave., East Orange, NJ 07017. All.
*Water Distribution Operator Training Handbook.* AWWA (1976). pp. IX-1 to IX-20.
Linsley, R.K. & Franzini, J.B. *Water Resource Engineering.* McGraw-Hill Book Company (1964). pp. 340-357.

**Meters**

*Stevens Water Resources Data Book.* Leupold and Stevens, Inc., Beaverton, OR (lst ed.). pp. 21, 37-47, 108, 109-111, 115.
*Water Distribution Operator Training Handbook.* AWWA (1976). pp. X-1 to X-12.

## Glossary Terms Introduced in Module 4

(Terms are defined in the Glossary at the back of the book.)

Aqueduct
Axial-flow pump
Bowl
Bowl discharge
Broad-crested weir
Bypass meter

Casing
Cathodic protection
Centrifugal pump
Conduit
Crest (of a weir)
Diffuser pump

Discharge
Discharge column
Electrolysis
Eye (of an impeller)
Flow-nozzle meter
Flume

Head
Head loss
Impeller
Intake structure
Line shaft
Magnetic meter

Malleable
Metering flume
Mixed-flow pump
Multistage pump
Nodule
Orifice meter

Parshall flume
Pipeline
Potable
Pressure-differential
meters
Priming

Propeller meter
Proportional meter
Radial-flow pump
Rectangular weir
Shaft
Sharp-crested weir

Single-stage pump
Staff gage
Stilling well
Suction
Surge tank
Throttle

Transmission line
Tuberculation
Turbine meter
Turbine pump
V-notch weir
Venturi meter

Vertical turbine pump
Volute pump
Water distribution
Water hammer
Water transmission
Weir

# Review Questions

(Answers to Review Questions are given at the back of the book.)

1. List and describe three types of intake structures.

2. Define aqueduct and list three types.

3. (a) List at least three types of pipe that are used in water transmission.
   (b) What is an advantage and disadvantage of each type of pipe listed above?

4. List at least three typical pipe joints and describe how each joint is assembled.

5. List five uses of valves in transmission lines.

6. Explain why this statement is true or false:
   *When a butterfly valve is open, it offers low head loss.*

7. Why are some plug valves provided with multiple ports?

8. What is one of the uses of pressure relief valves?

9. What type of valve is installed at pipeline high points and what does it do?

10. What type of valve in installed at pipeline low points and what does it do?

11. Why should valves never be slammed shut or closed quickly?

12. What could happen if an air and vacuum relief valve stopped operating?

13. Using the figure shown, fill in the number or numbers which represent the following terms:

    (a) _____ Static suction head
    (b) _____ Piezometric surface
    (c) _____ Static discharge head
    (d) _____ Total dynamic head
    (e) _____ Total static head
    (f) _____ Dynamic suction head
    (g) _____ Head loss
    (h) _____ Dynamic discharge head

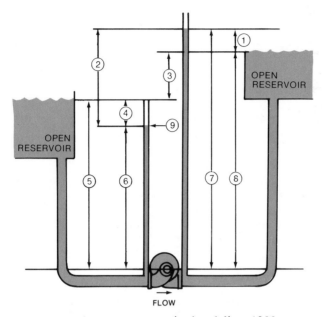

FLOW

14.  Calculate the water horsepower required to deliver 1200 gpm at a total head of 240 ft.

15.  Using the pump curve below, identify the total head, pump efficiency, and horsepower requirements for a pump operating at 4500 gpm.

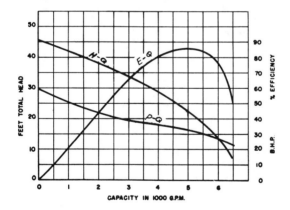

16.  Assuming a motor efficiency of 85 percent, calculate the motor horsepower required to produce 56 brake horsepower (Bhp).

17.  If electricity costs 2.5 cents per kilowatt hour, how much will it cost per hour to operate the above pump and motor when pumping at 1400 gpm against a head of 135 ft?

18. Name the type of pumps used routinely in water transmission.

19. If a centrifugal pump has one impeller and matching casing it's called a
    _____.

20. What are four advantages of using a centrifugal pump?

21. Which of the following are disadvantages of using a centrifugal pump?
    (a) Pumps produce a steady, uniform discharge.
    (b) The maximum suction lift (distance from the impeller shaft down to
        the water being pumped) is usually about 12 ft.
    (c) Pumps are not self-priming without additional equipment.
    (d) Air leaks on suction side reduce performance.

22. A mixed flow pump combines the operating principles of what other two
    pumps?

23. List two primary applications of vertical turbine pumps.

24. Why is a check valve important on the discharge side of a pump?

25. There are two general types of water meters—those that measure high flow
    rates and those that measure low flow rates. Where might a low flow meter
    be used?

26. Explain why the following statement is true or false:
    *A V-notch weir is the same thing as a 90-deg V-notch weir.*

27. If the notch of a rectangular weir were as wide as the channel, the weir
    would be called a _____.

28. Why is the depth of water (from the crest of a weir to the water surface)
    always measured some distance in back of the weir?

29. The head on a 90-deg V-notch weir is 1.89 ft. Using Table 15, determine the
    flow rate over this weir, in mgd.

30. List two advantages of the Parshall flume over weirs.

31. A Parshall flume with a 12-in. throat shows a water depth at the staff gage
    of 0.54 ft. From Table 16, what is the flow rate in mgd?

32. What is the operating principle of a pressure differential meter?

33. The Venturi meter illustrated in Figure 103 was used to measure a flow

rate. The high pressure reading was 10 psig and the low pressure reading was 5 psig. The diameter of the Venturi at the high pressure tap was 2.0 ft and at the low pressure tap it was 0.68 ft. What is the flow rate in cfs? (Use the Venturi nomograph in *Basic Science Concepts and Applications,* Hydraulics Section, Flow Rate Problems.)

## Study Problems and Exercises

1. Prepare a map of your water system's raw water transmission line. Label the reaches according to gravity or pressure flow, identify the pipe material and couplings used, and list the various pressure ratings of the pipe. Show all information directly on the map.

2. Using the map prepared in Problem 1, locate and label all valves, listing type of valve, size, and purpose.

3. Continuing with Problem 1, locate all pump and booster stations. Identify the types of pumps and their rated capacities.

4. Select one pipe material, investigate it, and prepare a report on its suitability for use as a water transmission line. Give particular emphasis to the requirements for maintenance, repair, and replacement.

5. Select one valve from those discussed in this module, investigate it, and prepare a detailed list of routine maintenance requirements, identifying the frequency for each item of maintenance. (Valve manufacturers may be able to provide detailed information.)

6. Investigate a centrifugal pump and prepare a maintenance schedule.

7. Explain how a Venturi meter works, using information in the Hydraulics Section of *Basic Science Concepts and Applications.* Show how the flow rate is calculated.

# Appendix A

## Interim Primary Regulations
### and
### Secondary Guidelines
### of the
# Safe Drinking Water Act

# Table A-1.
# Summary of the
# Interim Primary Regulations

Maximum Contaminant Levels

| Type of Contaminant (Community Systems) | Type of Contaminant (Non-Community Systems) | | Maximum Contaminant Levels (MCLs) |
|---|---|---|---|
| Inorganic Chemicals All Water Systems ** | Inorganic Chemicals All Water Systems—** *Nitrate only*<br>(all other contaminants at state option) | • Arsenic | 0.05 mg/l |
| | | • Barium | 1. mg/l |
| | | • Cadmium | 0.010 mg/l |
| | | • Chromium | 0.05 mg/l |
| | | • Lead | 0.05 mg/l |
| | | • Mercury | 0.002 mg/l |
| | | • Selenium | 0.01 mg/l |
| | | • Silver | 0.05 mg/l |
| | | • Fluoride<br>(Annual average of maximum daily air temperatures.) | |
| | | a) 53.7F & below | 2.4 mg/l |
| | | b) 53.8-58.3F | 2.2 mg/l |
| | | c) 58.4-63.8F | 2.0 mg/l |
| | | d) 63.9-70.6F | 1.8 mg/l |
| | | e) 70.7-79.2F | 1.6 mg/l |
| | | f) 79.3-90.0F | 1.4 mg/l |
| | | • Nitrate (as N) | 10. mg/l |
| Organic Chemicals*** Surface Water Systems Only | Organic Chemicals*** at State Option | • Endrin | 0.0002 mg/l |
| | | • Lindane | 0.004 mg/l |
| | | • Methoxychlor | 0.1 mg/l |
| | | • Toxaphene | 0.005 mg/l |
| | | • 2, 4-D | 0.1 mg/l |
| | | • 2, 4, 5-TP (Silvex) | 0.01 mg/l |

| Category | Standard |
|---|---|
| Turbidity Surface Water Systems Only | • 1 TU monthly average (5 TU monthly average may apply at state option) —OR— • 5 TU average of two consecutive days |
| Microbiological Contaminants All Water Systems ** | When using membrane filter test: • 1 colony/100 ml for the average of all monthly samples —OR— • 4 colonies/100 ml in more than one sample if less than 20 samples are collected per month —OR— • 4 colonies/100 ml in more than 5 per cent of the samples if 20 or more samples are examined per month. When using multiple-tube fermentation test: (10-ml portions) • Coliform shall not be present in more than 10 per cent of the portions per month, • Not more than one sample may have three or more portions positive when less than 20 samples are examined per month, or • Not more than 5 per cent of the samples may have three or more portions positive when 20 or more samples are examined per month. |
| Radiological Contaminants (Natural)— All Water Systems ** | • Gross Alpha    15 pCi/l • Combined Ra-226 and Ra-228    5 pCi/l |
| Radiological Contaminants (Man-made)— Surface water systems serving polulations greater than 100 000. | • Gross Beta    50 pCi/l • Tritium    20 000 pCi/l • Strontium-90    8 pCi/l |

*For all non-community water systems, *initial sampling and testing* must be conducted for nitrates. *Routing sampling and testing*, however, is *at state option.*

**Systems using surface and/or groundwater.

***Additional organic regulations will be issued. These proposed regulations will establish MCLs for the group of organics known as *trihalomethanes* and will specify treatment methods.

# Table A2.
# Summary of the
# Secondary Guidelines

| Contaminant | Level |
|---|---|
| Chloride | 250 mg/L |
| Color | 15 color units |
| Copper | 1 mg/L |
| Corrosivity | Non-corrosive |
| Foaming agents | 0.5 mg/L |
| Hydrogen sulfide | 0.05 mg/L |
| Iron | 0.3 mg/L |
| Manganese | 0.05 mg/L |
| Odor | 3 threshold odor number |
| pH | 6.5–8.5 |
| Sulfate | 250 mg/L |
| Total dissolved solids | 500 mg/L |
| Zinc | 5 mg/L |

# Appendix B

# Supplementary Data
for
Study Problems and Exercises

# Key to Using Supplementary Data

| Module | Study Problem and Exercise Number | Use Data From |
|---|---|---|
| 1 | 4 | Table B-1 |
| 1 | 6 | Table B-2 |
| 2 | 3 | Table B-3 |
| 2 | 4 | Table B-3 |
| 2 | 5 | Table B-4 |

**Table B-1.   Raw Water Quality Data    (January 1977—Daily values)**

| Day | Maximum Air Temperature °F | Turbidity NTU | Color CU | Odor TON | pH | Hardness as mg/L CaCO3 | TDS mg/L |
|---|---|---|---|---|---|---|---|
| 1 | 38.4 | 0.41 | 2 | 1 | 8.7 | 130 | 245 |
| 2 | 12.1 | 0.42 | 3 | 2 | 8.7 | 136 | 253 |
| 3 | 19.0 | 0.48 | 2 | 1 | 8.8 | 137 | 175 |
| 4 | 6.0 | 0.50 | 3 | 2 | 8.9 | 137 | 170 |
| 5 | 27.8 | 0.40 | 2 | 1 | 8.7 | 139 | 174 |
| 6 | 21.2 | 0.43 | 4 | 3 | 8.7 | 150 | 228 |
| 7 | 24.7 | 0.40 | 4 | 4 | 8.6 | 153 | 219 |
| 8 | 45.8 | 0.40 | 3 | 3 | 8.4 | 154 | 231 |
| 9 | 59.6 | 0.45 | 4 | 3 | 8.8 | 149 | 192 |
| 10 | 65.7 | 0.45 | 4 | 3 | 9.0 | 145 | 204 |
| 11 | 12.8 | 0.46 | 6 | 3 | 8.8 | 168 | 260 |
| 12 | 10.7 | 0.41 | 6 | 3 | 8.8 | 164 | 257 |
| 13 | 5.4 | 0.55 | 8 | 4 | 8.7 | 162 | 255 |
| 14 | 11.8 | 0.40 | 4 | 4 | 8.9 | 166 | 262 |
| 15 | 15.9 | 0.38 | 4 | 3 | 9.0 | 168 | 271 |
| 16 | 22.0 | 0.47 | 3 | 2 | 8.5 | 172 | 275 |
| 17 | 37.8 | 0.37 | 4 | 4 | 8.9 | 130 | 194 |
| 18 | 45.6 | 0.35 | 4 | 3 | 8.7 | 135 | 180 |
| 19 | 32.4 | 0.35 | 3 | 4 | 8.6 | 139 | 186 |
| 20 | 28.7 | 0.36 | 2 | 2 | 8.6 | 140 | 199 |
| 21 | 27.2 | 0.25 | 2 | 1 | 8.5 | 141 | 201 |
| 22 | 28.3 | 0.28 | 3 | 2 | 8.5 | 152 | 219 |
| 23 | 17.0 | 0.25 | 1 | 1 | 8.5 | 157 | 224 |
| 24 | 21.3 | 0.28 | 3 | 6 | 8.5 | 168 | 196 |
| 25 | 68.0 | 0.28 | 3 | 4 | 8.5 | 175 | 202 |
| 26 | 65.3 | 0.28 | 3 | 3 | 8.9 | 141 | 184 |
| 27 | 62.1 | 0.30 | 4 | 3 | 8.7 | 144 | 179 |
| 28 | 27.2 | 0.20 | 4 | 2 | 8.7 | 143 | 174 |
| 29 | 28.1 | 0.20 | 6 | 8 | 9.0 | 139 | 168 |
| 30 | 27.9 | 0.26 | 4 | 4 | 9.0 | 140 | 165 |
| 31 | 28.0 | 0.28 | 4 | 1 | 8.6 | 136 | 140 |

**Table B-2.   Treated Water Quality Data**

| Chemical Constituent | Constituent Concentrations (mg/L) | |
|---|---|---|
| | Jan. 26, 1977 | Oct. 11, 1977 |
| Arsenic | 0.01 | 0.001 |
| Barium | 0.5 | 0.1 |
| Cadmium | 0.0001 | 0.00021 |
| Lead | 0.002 | 0.0015 |
| Mercury | 0.0008 | 0.0004 |
| Selenium | 0.000 | 0.002 |
| Silver | 0.00002 | 0.0004 |
| Fluoride | 0.85 | 0.84 |
| Nitrate | 0.41 | 0.15 |

**Table B-3.   Volume of Water Treated[1]**
**1977 Annual Summary**

| Month | Total Volume Treated gallons | Highest Day gallons | Lowest Day gallons |
|---|---|---|---|
| January | 92,480,000 | 3,982,000 | 2,187,000 |
| February | 101,600,000 | 4,080,000 | 2,346,000 |
| March | 103,250,000 | 4,117,000 | 2,724,000 |
| April | 112,840,000 | 4,522,000 | 3,012,000 |
| May | 119,720,000 | 4,671,000 | 3,297,000 |
| June | 124,600,000 | 5,269,000 | 3,508,000 |
| July | 132,000,000 | 5,340,000 | 3,721,000 |
| August | 117,960,000 | 4,762,000 | 3,305,000 |
| September | 81,570,000 | 3,629,000 | 2,368,000 |
| October | 76,310,000 | 3,075,000 | 2,279,000 |
| November | 64,470,000 | 2,649,000 | 1,965,000 |
| December | 75,100,000 | 2,920,000 | 2,090,000 |

[1]Serving a population of 15,840

## Table B-4. Rates of Flow Measured Every Half Hour*

| Time of Day | Date of Flow Record | | | | |
|---|---|---|---|---|---|
| | 6/27/77 | 7/18/77 | 12/26/77 | 12/20/77 | 2/12/78 |
| 12 M | 9.00 | 11.05 | 4.95 | 4.95 | 3.85 |
| | 8.55 | 10.10 | 4.30 | 4.40 | 3.30 |
| 1 AM | 8.75 | 9.80 | 4.00 | 3.85 | 3.40 |
| | 8.90 | 9.85 | 3.60 | 3.65 | 2.70 |
| 2 | 8.40 | 9.75 | 3.35 | 3.30 | 2.60 |
| | 8.20 | 9.90 | 3.25 | 3.00 | 2.40 |
| 3 | 9.30 | 10.20 | 3.20 | 2.90 | 2.30 |
| | 9.35 | 10.65 | 2.95 | 2.60 | 2.20 |
| 4 | 10.80 | 11.20 | 2.80 | 2.70 | 2.10 |
| | 11.55 | 12.20 | 2.75 | 2.70 | 2.05 |
| 5 | 12.85 | 12.90 | 2.95 | 2.80 | 2.15 |
| | 13.00 | 13.50 | 3.30 | 2.95 | 2.25 |
| 6 | 15.60 | 16.00 | 3.70 | 3.25 | 2.50 |
| | 20.10 | 21.90 | 4.10 | 3.70 | 2.95 |
| 7 | 22.20 | 24.90 | 5.35 | 4.85 | 3.80 |
| | 24.60 | 27.60 | 7.00 | 6.35 | 5.40 |
| 8 | 26.00 | 28.40 | 8.60 | 7.55 | 6.80 |
| | 28.20 | 30.60 | 10.00 | 8.55 | 8.30 |
| 9 | 28.40 | 32.10 | 10.80 | 8.80 | 8.80 |
| | 28.80 | 31.70 | 11.75 | 9.35 | 9.65 |
| 10 | 30.40 | 31.70 | 11.60 | 9.80 | 10.50 |
| | 28.50 | 30.60 | 12.00 | 9.70 | 10.80 |
| 11 | 28.20 | 28.80 | 11.95 | 9.40 | 11.00 |
| | 27.60 | 29.40 | 11.55 | 9.20 | 10.80 |
| 12 N | 26.70 | 26.40 | 10.95 | 9.00 | 10.70 |
| | 25.80 | 26.70 | 10.80 | 8.95 | 10.20 |
| 1 PM | 26.40 | 26.40 | 10.60 | 8.40 | 9.70 |
| | 25.20 | 25.80 | 10.00 | 8.30 | 9.60 |
| 2 | 25.50 | 25.50 | 9.25 | 7.80 | 9.20 |
| | 24.60 | 25.50 | 9.05 | 7.10 | 8.65 |
| 3 | 24.00 | 24.00 | 9.00 | 7.25 | 8.40 |
| | 25.50 | 24.20 | 8.75 | 7.15 | 8.25 |
| 4 | 26.10 | 24.80 | 8.90 | 6.95 | 7.80 |
| | 28.50 | 26.70 | 9.20 | 6.90 | 8.80 |
| 5 | 32.10 | 29.10 | 9.80 | 7.00 | 8.80 |
| | 36.60 | 34.20 | 9.80 | 6.85 | 9.40 |
| 6 PM | 38.40 | 39.90 | 9.80 | 6.90 | 8.60 |
| | 40.80 | 45.00 | 9.35 | 6.60 | 8.90 |
| 7 | 42.00 | 47.70 | 8.80 | 7.00 | 9.60 |
| | 41.70 | 49.40 | 8.40 | 5.80 | 8.60 |
| 8 | 41.40 | 50.40 | 7.95 | 6.50 | 8.00 |
| | 41.70 | 50.00 | 7.10 | 5.20 | 7.40 |
| 9 | 40.60 | 47.20 | 7.00 | 6.20 | 8.40 |
| | 33.20 | 40.60 | 6.75 | 5.55 | 6.80 |
| 10 | 27.30 | 33.60 | 6.80 | 6.00 | 6.80 |
| | 19.60 | 26.40 | 6.65 | 6.00 | 7.70 |
| 11 | 14.40 | 18.60 | 7.00 | 6.00 | 4.90 |
| | 11.70 | 14.50 | 5.25 | 5.15 | 3.80 |

*This table shows flow of treated water produced by the plant. The flow from the plant may not follow the same pattern as the flow through the distribution system, due to movement of treated water from the plant into storage tanks and from storage tanks into distribution.

# Glossary

# Glossary

Words defined in the glossary are set in SMALL CAPITAL LETTERS where they are first used in the text.

**Acidic water, 18**   Water having a pH less than 7.0.

**Aeration, 20**   The process of bringing water and air into close contact to remove or modify constituents in the water.

**Aesthetic, 17**   Having to do with attractiveness to the senses.

**Air-lift pump, 105**   A pump, used largely for lifting water from wells. Air under pressure is discharged into the water at the bottom of the well in fine bubbles. The bubbles mix with the water and reduce the effective specific gravity of the air–water mixture, causing the mixture to rise in the discharge pipe to the outlet.

**Algae, 17**   Primitive plants, (one- or many-celled), that usually live in water and are capable of obtaining their food by PHOTOSYNTHESIS.

**Alkaline water, 18**   Water having a pH greater than 7.0. Also called BASIC WATER.

**Anaerobic, 26**   The absence of air or free oxygen.

**Annual average daily flow, 53**   The average of the DAILY FLOWS for a 12-month period. May also be found by dividing the total volume for the year by 365, the number of days in the year.

**Annular space, 70**   The space between the outside of a well CASING and the drilled hole.

**Appropriative rights, 10**   Water rights acquired by diverting and putting the water to beneficial use following procedures established by state statutes or courts.

**Aquatic life, 20**   All forms of animal and plant life that live in water.

**Aqueduct, 83**   A CONDUIT, usually of considerable size, used to convey water.

**Aquifer, 2,10**   A porous, water-bearing geologic formation. Generally restricted to materials capable of yielding an appreciable supply of water. Also called GROUND-WATER AQUIFER.

**Artesian aquifer, 12**   An AQUIFER in which the water is confined by both an upper and a lower IMPERMEABLE LAYER.

**Artesian well, 12**   A well in which the water rises above the upper confining or IMPERMEABLE LAYER. In a *flowing artesian well* the water will rise to the ground surface and flow out onto the ground. In a *non-flowing artesian well* the water rises above the upper impermeable layer but does not rise as high as the surface of the ground.

**Atom, 15**   The basic structural unit of matter; the smallest particle of an element that can combine chemically with similar particles of the same or other elements to form MOLECULES of a compound.

**Average daily flow, 53**   The sum of all DAILY FLOWS for a specified time period divided by the number of DAILY FLOWS added.

**Axial-flow pump, 107**   A pump in which a propeller-like IMPELLER forces water out in a direction parallel to the SHAFT.

**Bacteria, 23**   A group of one-celled microscopic organisms that have no chlorophyll. Bacteria usually have spherical, rod-like, or curved shapes. Usually regarded as plants.

**Basic water, 18**   Water having a pH greater than 7. See ALKALINE WATER.

**Bowl, 108**   The CASING surrounding a TURBINE-PUMP IMPELLER.

**Bowl discharge, 108**   In a VERTICAL TURBINE PUMP , the port or ports in the IMPELLER casing (BOWL) through which water is discharged, either directly into the next stage of the pump or from the final stage into the pump discharge pipe.

**Broad-crested weir, 115**   A WEIR having a very wide crest over which the water must flow. Also called wide-crested weir.

**Bypass meter, 114**   Any flow meter that diverts a small portion of the main flow and measures the FLOW RATE of that portion as an indication of the rate of the main flow. Also called PROPORTIONAL METER, since the rate of the diverted flow is proportional to the rate of the main flow.

**Capillary action, 2**   The characteristic that describes how a material or an object, containing minute openings or passages, when partly immersed in water, will draw water up into the openings or passages to a level above the free water surface.

**Casing, 70,104**   (1) The enclosure surrounding a pump IMPELLER, into which are machined the suction and discharge ports. (2) The metal pipe used to line the borehole of a well. Also called WELL CASING.

**Cathodic protection, 85**   An electrical system for prevention of corrosion to metals, particularly metallic pipe.

**Centrifugal pump, 103**   A pump consisting of an IMPELLER on a rotating SHAFT which is in a CASING having suction and a discharge connection. The rotating impeller creates pressure by the velocity derived from centrifugal force.

**Coliform-group bacteria, 23**   A group of BACTERIA predominantly inhabiting the intestines of man or animal, but also occasionally found elsewhere. Presence of the bacteria in the water is used as an indication of fecal contamination (contamination by human or animal wastes).

**Color, 17**   A physical characteristic describing the apperance of water (different from TURBIDITY, which is the cloudiness of water).

**Color unit (CU), 17**   The unit of measure used to express the color of a water sample.

**Commercial water use, 46**   Use of water in business activities such as motels, shopping centers, gas stations, and laundries.

**Condensation, 2**   The process by which a substance changes from the gaseous form to a liquid or solid form. Water that falls as PRECIPITATION from the atmosphere has condensed from the vapor (gaseous) state to rain or snow. Dew and frost are also forms of condensation on the surface of the earth or vegetation.

**Conduit, 83**   A tube or channel used to convey water or any other fluid.

**Cone of depression, 71**   The cone-shaped depression in the GROUND-WATER level around a well during pumping.

**Contamination, 28**   Any introduction into water of microorganisms, chemicals, wastes, or wastewater in a concentration that makes the water unfit for its intended use.

**Corrosion, 19**   The gradual deterioration or destruction of a substance or material by chemical action. The action proceeds inward from the surface.

**Crest (of a weir), 115**   The top of a dam, dike, spillway, or WEIR.

**Daily flow, 53**   The total number of gallons of water passing through a plant during a 24-hour period.

**Demand factor, 55**   The ratio of the peak or minimum demand to the average demand.

**Diffuser, 107**    A pump that uses vanes mounted within the CASING to change the high velocity, low pressure flow leaving the impeller to a lower velocity, higher pressure flow at the discharge.

**Discharge, 104**    The hole, or port, in a pump CASING through which the water exits.

**Discharge column, 108**    The pipe used to suspend and support a well pump assembly. The pipe also carries the ground water to the surface. Also called a drop pipe.

**Displacement pump, 105**    A type of pump in which the water is forced to flow from the source of supply through an inlet pipe and inlet valve and into the pump chamber by a vacuum. The vacuum is created by the withdrawal of a piston or piston-like device that, on its return, displaces a certain volume of the water contained in the chamber and forces it to flow through the discharge valves and discharge pipes.

**Dissolved oxygen (DO), 20**    The oxygen dissolved in water, wastewater, or other liquid, usually expressed in MILLIGRAMS PER LITER, PARTS PER MILLION, or percent of saturation.

**Dissolved solids, 21**    Any material that is dissolved in water and can be recovered by evaporating the water after filtering the suspended material and measuring the weight of the residue remaining.

**Distribution system, 79**    The network of pipes, commonly in a grid arrangement, through which treated water is delivered to consumers.

**Divide, 6**    The line that follows the ridges or summits forming the boundary of a DRAINAGE BASIN (watershed), separating one drainage basin from another. Also called WATERSHED DIVIDE.

**Domestic water use, 45**    Use of water for household purposes such as washing, drinking, flushing, and lawn watering.

**Drainage basin, 6**    An area from which SURFACE RUNOFF is carried away by a single drainage system. Also called catchment area, WATERSHED, drainage area.

**Drainage divide, 6**    See DIVIDE.

**Drawdown, 70**    The difference between the STATIC WATER LEVEL and the PUMPING WATER LEVEL in a well.

**Effluent, 24**    Water flowing out of a structure such as a treatment plant.

**Ejector pump, 105**    A pump for moving a fluid by drawing it along in a high-velocity stream of air or water.

**Electrical conductivity (EC), 22**    A test that measures the ability of water to transmit electricity. Electrical conductivity is an indicator of DISSOLVED SOLIDS concentration. Normally an EC of 1000 $\mu$mhos per square centimeter indicates a dissolved solids concentration of 600–700 mg/L.

**Electrolysis, 91**    The corrosion of metals by an electric current.

**Ephemeral stream, 8**    (1) A stream that flows only in direct response to PRECIPITATION. Such a stream receives no water from springs and no continued supply from melting snow or other surface source. Its channel is above the water table at all times. (2) Streams or stretches of streams that do not flow continuously during periods of as much as one month.

**Evaporation, 2**    The process by which water becomes a vapor at a temperature below the boiling point. The rate of evaporation is generally expressed in inches or centimeters per day, month, or year.

**Eye (of an impeller), 106**    The center of an IMPELLER.

**Filtrable residue test, 22**    A test used to measure the total dissolved solids in water by first filtering out any undissolved solids and then evaporating the filtered water to dryness. The residue that remains is called filtrable residue or total dissolved solids.

**Flow, 51**    General term for movement of water, commonly used to mean (imprecisely) INSTANTANEOUS FLOW RATE, average flow rate, or VOLUME.

**Flow-nozzle meter, 112**   A water meter of the PRESSURE-DIFFERENTIAL type in which the FLOW through a nozzle produces a pressure difference which a float tube then uses as an indication of the rate of flow.

**Flow rate, 50**   The volume of water passing by a point per unit of time. Flow rates are either INSTANTANEOUS or average.

**Flume, 117**   An open CONDUIT of wood, masonry, or metal constructed on a grade and usually elevated. See also METERING FLUME.

**Gallons per capita per day (gpcd), 44**   The number of gallons used by one person in one day.

**Gravel packing, 70**   Gravel surrounding the WELL INTAKE SCREEN, artificially placed ("packed") to aid the screen in filtering out the sand of the aquifer. Only needed in AQUIFERS containing a large proportion of fine-grained material.

**Ground water, 2**   Subsurface water occupying the SATURATION zone, from which wells and springs are fed. In a strict sense the term applies only to water below the WATER TABLE. Also called phreatic water.

**Ground-water aquifer, 2,10**   See AQUIFER.

**Grout, 70**   As specifically used in sealing water wells, grout is a mixture of portland cement and water, usually 5 to 6 gallons of water mixed with one 94-lb sack of cement. Sometimes clay or sand is added. Clay helps to hold the cement in suspension, reduces shrinkage, and allows the grout to flow more smoothly. Sand or other bulky materials can be added to help the grout bridge larger openings in the formation being sealed. *Grouting* is the act of placing the grout in the space between the well casing and the hole.

**Hardness, 19**   A chracteristic of water, caused primarily by the salts of calcium and magnesium. Causes deposition of scale in boilers, damage in some industrial processes, and sometimes objectionable taste. May also decrease the effectiveness of soap.

**Head, 96**   (1) A measure of the energy possessed by water at any particular location in the water system, expressed in feet. (2) A measure of the pressure or force exerted by water, expressed in feet.

**Head loss, 96**   The amount of energy used by water in moving from one point to another.

**Impeller, 104**   The rotating set of vanes that forces water through a pump.

**Impermeable layer, 11**   A layer not allowing, or allowing only with great difficulty, the movement of water.

**Impoundment, 63**   A pond, lake, tank, basin, or other space, either natural or man-made, which is used for storage, regulation, and control of water.

**Industrial water use, 46**   Use of water in industrial activities such as power generation, steel manufacturing, beverage manufacturing, pulp and paper processing, and food processing.

**Infiltration, 2**   The flow or movement of water through soil.

**Influent**   Water flowing into a structure such as a water treatment plant.

**Inorganic material (inorganics), 16**   Chemical substances of mineral origin, or more correctly, not of basically carbon structure.

**Instantaneous flow rate, 51**   The FLOW RATE (volume per time) at any instant. Defined by the equation $Q = AV$, where $Q$ = flow rate; $A$ = cross-sectional area of the water; and $V$ = velocity.

**Intake structure, 65,80**   A structure or device placed in a surface water source to permit the withdrawal of water from that source.

**Jet pump**   A device that pumps fluid by converting the energy of a high-pressure fluid into that of a high-velocity fluid.

**Line shaft, 108**   The drive shaft connecting a submerged TURBINE PUMP assembly with the motor assembly mounted at the well head.

**Magnetic meter, 113**    A FLOW measuring device in which the movement of water induces an electrical current proportional to the rate of flow.

**Malleable, 84**    The characteristic that describes any material that can be hammered, pounded, or pressed into various shapes without breaking or returning to its original shape.

**Maximum contaminant level (MCL), 28**    The maximum permissible level of a CONTAMINANT in water as specified in the regulations of the Safe Drinking Water Act.

**Metering flume, 117**    A FLOW measuring device such as a PARSHALL FLUME that is used to measure flow in an open channel.

**Milligrams per litre (mg/L), 15**    A unit of the concentration of water or wastewater constituent. It is 0.001 gram of the constituent in 1000 millilitres of water. In reporting the results of water and wastewater analyses it has replaced the unit formerly used commonly, PARTS PER MILLION , to which it is approximately equivalent.

**Millirem, 33**    An expression or measure of the extent of biological injury that would result from the absorption of a particular RADIONUCLIDE at a given dosage over one year.

**Minimum day, 55**    The day during which the MINIMUM DAY DEMAND occurs.

**Minimum day demand, 55**    Least volume per day flowing through the plant for any day of the year.

**Minimum hour demand, 55**    Least volume per hour flowing through a plant for any hour in the year.

**Minimum month, 55**    The month during which the MINIMUM MONTH DEMAND occurs.

**Minimum month demand, 55**    Least volume of water passing through the plant during a calendar month.

**Mixed flow pump, 107**    A pump that moves water partly by centrifugal force and partly by the lift of vanes on the liquid. With this type of pump the flow enters axially and leaves axially and radially.

**Molecule, 15**    The smallest division of matter that still retains the characteristics of the original material. If the original material is an element, the molecule may be a single ATOM; if the original material is a chemical compound, the molecule will be composed of two or more different atoms, chemically bonded together.

**Multistage pump, 108**    A pump that has two or more IMPELLERS and matching CASINGS.

**Nodule, 86**    A small knob or button of rust or metallic salts, found on the inside of pipe. Also called a *tubercule*.

**Organic material (organics), 16**    Chemical substances of animal or vegetable origin, or more correctly, of basically carbon structure.

**Orifice meter, 112**    A type of FLOW meter consisting of a section of pipe blocked by a disc with a small hole, or orifice. The entire flow passes through the orifice, creating a pressure drop proportional to the flow rate.

**Oxidize, 24**    To chemically combine with oxygen. In the case of BACTERIA or other organics, the combination of oxygen and organic material to form more stable organics or minerals.

**Palatable (palatability), 1**    Pleasing to the taste.

**Parshall flume, 119**    A calibrated channel for measuring the FLOW of liquid in an open conduit.

**Parts per million (ppm), 15**    The number of weight or volume units of a constituent present with each one million units of the solution or mixture. Formerly used to express the results of most water and wastewater analyses but recently replaced by the ratio MILLIGRAMS PER LITRE.

**Pathogens (pathogenic), 23**    Disease-causing organisms.

**pCi (picocurie), 33**    A measure of the disintegration of a particular RADIONUCLIDE.

**pCi/L (picocurie per litre), 33** A measure of the concentraion of a particular RADIONUCLIDE based on its disintegration or decay rate.

**Peak day, 54** The day during which the PEAK DAY DEMAND occurs.

**Peak day demand, 54** Greatest volume per day flowing through a plant for any day of the year.

**Peak hour demand, 55** Greatest volume per hour flowing through a plant for any hour in the year.

**Peak month, 55** The month during which the PEAK MONTH DEMAND occurs.

**Peak month demand, 55** The greatest volume of water passing through the plant during a calendar month.

**Percolation, 2** The movement or flow of water through the pores of soil, usually downward.

**Perennial stream, 8** A stream that flows continuously at all seasons of a year and during dry as well as wet years. Such a stream is usually fed by GROUND WATER , and its water surface generally stands at a lower level than that of the WATER TABLE in the locality.

**Permeability, 68** (1) The property of a material that permits movement of water through the material. (2) A measure of how easily water flows through a material.

**pH, 18** A measure of the acid or alkaline value of water. pH values range from 1 (most acidic) to 14 (most alkaline); pure water, which is neutral (neither acidic nor alkaline), has a pH of 7, the center of the range.

**Photosynthesis, 20** The process by which plants, using the chemical CHLOROPHYLL, convert the energy of the sun into food energy. Through photosynthesis, all plants, and ultimately all animals (which feed on plants or other, plant-eating animals), obtain the energy of life from sunlight.

**Piezometric surface, 13** The surface that coincides with the STATIC WATER LEVEL in an ARTESIAN AQUIFER.

**Pipeline, 84** Lengths of pipe joined together to provide a conduit through which fluids flow.

**Pollution, 17** A condition created by the presence of harmful or objectionable material in water.

**Potable** The characteristic that describes water that does not contain objectional POLLUTION, CONTAMINATION, minerals, or infective agents and is considered satisfactory for domestic consumption.

**Precipitation, 2** The process by which atmospheric moisture is discharged onto land or water surface. Precipitation, in the form of rain, snow, hail, or sleet, is usually expressed as depth in a day, month, or year, and designated as daily, monthly, or annual precipitation.

**Pressure-differential meters, 112** Any FLOW measuring device that creates and measures a difference in pressure proportionate to the rate of flow. Examples include the VENTURI METER, ORIFICE METER, and FLOW NOZZLE.

**Priming, 110** The action of starting the flow in a pump or siphon.

**Propeller meter, 114** A meter for measuring FLOW RATE by measuring the speed at which a propeller spins in, and hence the velocity at which the water is moving through, a CONDUIT of known cross-sectional area.

**Proportional meter, 114** See BYPASS METER.

**Protozoa, 23** Small single-celled animals including amoebae, ciliates, and flagellants.

**Public water use, 47** Use of water to support public facilities such as public buildings and offices, parks, golf courses, picnic grounds, camp grounds, and ornamental fountains.

**Pump, 103** See type; CENTRIFUGAL, DISPLACEMENT, DIFFUSER, AXIAL FLOW, RADIAL FLOW, MIXED FLOW, VERTICAL TURBINE, REGENERATIVE TURBINE.

**Pumping water level, 70** The distance from the ground surface to GROUND-WATER level when pumping is in progress.

**Radial-flow pump, 106**     A pump that moves water by centrifugal force, spinning the water radially outward from the center of the impeller. The DIFFUSER and the VOLUTE centrifugal pumps are commonly used in water utilities.

**Radioactive material, 23**     A material with an unstable atomic nucleus, which spontaneously decays or disintegrates, producing radiation.

**Radionuclide, 23**     See RADIOACTIVE MATERIALS.

**Radius of influence, 71**     The distance from the center of a well to the outermost edge of the CONE OF DEPRESSION.

**Rainfall intensity, 5**     The amount of rainfall occurring during a specified time. Rainfall intensity is usually expressed as inches or centimeters per hour.

**Recharge, 5**     Addition of water to the GROUND-WATER supply from PRECIPITATION and by INFILTRATION from surface streams, lakes, reservoirs, and snow melt.

**Reciprocating displacement pump, 105**     A type of pump consisting essentially of a closed cylinder containing a piston or plunger as the displacing mechanism, which draws liquid into the cylinder through an inlet valve and forces it out through an outlet valve.

**Rectangular weir, 115**     A WEIR having a notch that is rectangular.

**Regenerative turbine, 105**     A hydraulic mechanism, used in hydroelectric dams, automobile automatic transmissions, jet aircraft engines, and in special pump designs.

**Residual drawdown, 72**     The distance between original STATIC WATER LEVEL and the water level after pumping has stopped.

**Riparian rights, 9**     A water right that assures the owner of land abutting a stream or other natural body of water the use of that water.

**Runoff, 4**     See SURFACE RUNOFF.

**Safe yield**     The maximum dependable water supply that can be withdrawn continuously from a SURFACE or GROUND-WATER supply during a period of years in which the driest period or period of greatest deficiency in water supply is likely to occur.

**Sanitary seal, 70**     A cap that prevents contaminants from entering a well through the top of the CASING.

**Saturated, 4**     A material is saturated with a second material when it can absorb no more of the second material. Saturated soil has its VOID SPACES completely filled with water, so any water added will run off and not soak in.

**Screw-feed pump**     A pump with either horizontal or vertical cylindrical CASING, in which operates a runner with radial blades like those of a ship's propeller.

**Seepage, 66**     The slow movement of water through small cracks or pores of a material.

**Shaft, 104**     The bearing-supported rod in a pump turned by the motor, on which the IMPELLER is mounted.

**Sharp-crested weir, 116**     A WEIR having a crest, usually consisting of a thin metal plate, so sharp that the water, in passing over it, touches only the line of the crest edge.

**Silting, 66**     The accumulation of silt (small soil particles between 0.004 and 0.062 mm diameter) in an impoundment.

**Single-stage pump, 108**     A pump that has one IMPELLER and matching CASING.

**Specific capacity, 72**     A measure of WELL YIELD per unit of DRAWDOWN.

**Staff gage, 116**     A graduated scale on a plank, metal plate, pier, wall, etc., used to indicate the height of a fluid level.

**Static water level, 70**     The water level in a well when no water is being taken from the AQUIFER.

**Stilling well, 116**     A pipe, chamber, or compartment with comparatively small inlet or inlets connecting it with a main body of water. Its purpose is to dampen waves or surges while permitting the water level within the well to rise and fall with the major fluctuations of the main body of water. It is used with water-measuring devices to improve accuracy of measurement.

**Suction, 104**     The hole, or port, in a pump CASING into which water to be pumped enters.

**Surface runoff, 4** (1) That portion of the RUNOFF of a DRAINAGE BASIN that has not PERCOLATED beneath the surface after PRECIPITATION. (2) The water that reaches a stream by traveling over the soil surface or by falling directly into the stream channels, including not only the large permanent streams but also the tiny rills and rivulets.

**Surface water, 4** All water on the surface, as distinguished from subsurface or GROUND WATER.

**Surge tank, 100** A tank or chamber used in large pipelines to absorb or cushion the increases in pressure caused by sudden changes in flow velocity, thus preventing WATER HAMMER.

**Temperature, 16** A physical characteristic of water. The temperature of water is normally measured on one of two scales; Fahrenheit or Celsius.

**Threshold odor, 18** The minimum odor of a water sample that can just be detected after successive dilutions with odorless water. Also called odor threshold.

**Throttle, 95** To open or close a valve to control the rate of FLOW. Throttling is normally used to describe the closing of a valve.

**Transmission line, 79** The PIPELINE or AQUEDUCT used for WATER TRANSMISSION.

**Transpiration, 2** The process by which water vapor is lost to the atmosphere from living plants.

**Tremie pipe** A pipe used for depositing concrete or gravel during well construction.

**Tuberculation, 85** The growth of nodules on the pipe interior; this reduces the inside diameter and increases the pipe roughness.

**Turbidity, 16** A physical characteristic of water making the water appear cloudy. The condition is caused by the presence of suspended matter.

**Turbidity unit (TU), 16** The unit of measure used to express the turbidity (cloudiness) of a water sample.

**Turbine meter, 114** A meter for measuring FLOW RATE by measuring the speed at which a turbine spins in water, indicating the velocity at which the water is moving through a CONDUIT of known cross-sectional area.

**Turbine pump, 107** (1) See DIFFUSER PUMP. (2) A REGENERATIVE TURBINE pump.

**Turnover, 16** The vertical circulation of water in large water bodies caused by the mixing effects of temperature and wind.

**V-notch weir, 115** A WEIR with a triangular opening.

**Venturi meter, 112** A PRESSURE DIFFERENTIAL METER for measuring FLOW of water or other fluid through closed CONDUITS or pipes, consisting of a venturi tube and a flow-registering device. The difference in velocity head between the entrance and the contracted throat of the tube is an indication of the rate of flow.

**Vertical turbine pump, 108** A CENTRIFUGAL PUMP, commonly of the MULTISTAGE, DIFFUSER type, in which the pump SHAFT is mounted vertically.

**Virus, 23** The smallest and simplest form of life. A virus (there are many types) reproduces itself in a manner that causes infectious disease in some larger life forms, such as man.

**Void space, 6** A pore or open space in rock or granular material that is not occupied by solid matter. It may be occupied by air, water, or other gaseous or liquid material. Also called interstice, or void.

**Volume, 50** Basic measurement of amount of water, such as gallons, cubic feet, cubic metres, litres, etc.

**Volute pump, 106** A type of CENTRIFUGAL PUMP in which the high velocity, low pressure water leaving the impeller is changed to lower velocity, higher pressure water by the expanding spiral shape (volute) of the casing.

**Water body, 6** Water impounded in natural or artificial basin.

**Waterborne disease, 32** A disease caused by a waterborne organism or toxic substance.

**Water course, 6**   (1) A running stream of water. (2) A natural or artificial channel for the passage of water.

**Water distribution, 79**   The process of delivering treated water to consumers.

**Water hammer, 100**   The potentially damaging slam, bang, or shudder in a pipe that occurs when a sudden change in water velocity (usually the result of too-rapidly starting a pump or operating of a valve) creates a great increase in water pressure.

**Watershed, 6**   See DRAINAGE BASIN.

**Watershed divide, 6**   See DIVIDE.

**Water table, 11**   The upper surface of the zone of SATURATION closest to the ground surface.

**Water table aquifer, 12**   An AQUIFER confined only by a lower IMPERMEABLE LAYER.

**Water table well, 12**   A well constructed in a WATER TABLE AQUIFER.

**Water transmission, 79**   The movement of water from its source to a point of storage, treatment, or distribution, or from the point of treatment to the WATER DISTRIBUTION system.

**Weir, 115**   An obstruction to the flow of water, placed in an open channel to measure FLOW RATE by measuring the depth of the water as it flows through a precisely sized and shaped notch in the weir.

**Well screen (intake screen), 70**   A specially designed screen placed below the PUMPING WATER LEVEL, through which the water must pass to reach the pump intake.

**Well slab, 70**   A concrete platform set on or into the ground around the CASING at the top of a well.

**Well yield, 72**   The FLOW RATE at which a well will discharge water on a sustained basis.

# Answers
# to
# Review Questions

# Answers to Review Questions

## Module 1   Sources and Characteristics

1. The constant movement of water from the earth to the atmosphere and back again.

2. The seven major elements which comprise the water cycle are: (1) evaporation, (2) transpiration, (3) condensation, (4) precipitation, (5) infiltration, (6) percolation, and (7) runoff.

3. Condensation.

4. Infiltration is the flow of water into the soil. Percolation is the deep, vertical movement of water between the soil and the ground-water aquifer.

5. Infiltration.

6. The following are some of the factors that influence the amount of surface runoff: rainfall intensity, length of rainfall, soil composition, soil moisture, slope of ground, vegetation covering the ground, and man-made influences.

7. When clay becomes wet it begins to swell. This swelling closes the void spaces in the soil, thus allowing less water to enter.

8. Watershed divide.

9. Perennial and ephemeral.

10. Precipitation, ground water seepage, springs, and snowmelt.

11. Riparian and appropriative.

12. The water table.

13. An aquifer is an underground layer of gravel, sand, sandstone, shattered rock, or limestone, or other formation that holds ground water.

14. (1) Water table aquifer, (2) Artesian aquifer.

15. (1) Physical, (2) Chemical, (3) Biological, (4) Radiological.

16. Milligrams per litre, abbreviated mg/L. This is preferred to parts per million (ppm).

17. Temperature can affect water treatment operations. For example, chemicals will dissolve easier in warm water than in cold water. Warmer temperatures tend to encourage the growth of various forms of plant life in waters. Generally, the consumer prefers the taste and cooling efficiency of colder water.

18. (a) $212°F$, (b) $14°F$, (c) $0°C$, (d) $20°C$, (e) $4°C$. (See *Basic Science Concepts and Applications,* Mathematics Section, Conversions, Temperature Conversions.) Water boils at $100°C$ ($212°F$). Water freezes at $0°C$ ($32°F$). Water is most dense at $4°C$ ($39.2°F$).

19. Turbidity.

20. a, b, e

21. Scaling of pipes, water heaters, meters, and boilers; increased cost of heating water due to scale formation; increased soap consumption; objectionable taste.

22. (a) 20, (b) 12.2, (c) 50. (See *Basic Science Concepts and Applications,* Chemistry Section, Equivalent Weights.)

23. 300 mg/L as $CaCO_3$. (See *Basic Science Concepts and Applications,* Chemistry Section, Equivalent Weights.)

24. Micromhos per centimeter at $25°C$, abbreviated $\mu$mhos/cm at $25°C$.

25. 420 to 490 mg/L. (See *Basic Science Concepts and Applications,* Chemistry Section, Laboratory Test Calculations—Electrical Conductivity.)

26. (1) Slow moving water allows suspended matter to settle.
    (2) Long holding time allows bacteria to oxidize non-settleable solids, converting them to settleable solids thus reducing color, taste and odor.
    (3) Natural aeration provides oxygen for a balanced population of aquatic life.
    (4) Large volume dilutes pollutants.

27. Yes, there is a risk. As shown in Table 5, animal and human feces are the sources of the disease-causing organisms. Whenever there is the possibility of fecal contamination of a water, there is the possibility that drinking the untreated water will cause disease.

28. Turnover is the vertical circulation that occurs in large bodies of water. Circulation can be caused by the mixing action of the wind and by the change in the density of water as its temperature becomes greater or less than $39.2°F$, the temperature of maximum water density. In the fall of the year, the surface waters cool and sink through the lighter lower waters, producing vertical circulation. In the spring of the year, the surface waters warm above the freezing point, increase in density and also tend to sink. The circulation stirs up the anaerobic bottom sediments, which degrades water quality, and produces obnoxious odors. The effects of turnover can last from several days to several weeks. This unique property of water together with the mixing action of water causes lake turnover.

29. $39.2°F$ or $4°C$

30. False. Ground water quality is very uniform and very slow to change.

31. Iron, manganese, fluoride, chloride, sulfate, nitrate, sodium, magnesium (hardness), calcium (hardness), and hydrogen sulfide.

32. *Turbidity* interferes with chlorination by providing hiding places for organisms where the chlorine cannot reach.

33. Arsenic, barium, cadmium, chromium, lead, mercury, selenium, silver, fluoride, nitrate, pesticides and herbicides.

34. 44.3 mg/L (about 45 mg/L) as $NO_3$. (See *Basic Science Concepts and Applications,* Chemistry Section, Equivalent Weights.)

35. The warmer the climate, the more water the consumer drinks. This is important because the adverse effects of fluoride depend on how much fluoride is taken into the body.

36. (a) Average annual temperature = 75.5° F (See *Basic Science Concepts and Applications,* Mathematics Section, Averages.)
    (b) Fluoride MCL = 1.6 mg/L.

37. Gastroenteritis, typhoid, bacillary dysentery, cholera, infectious hepatitis, amebic dysentery, and giardiasis.

# Module 2   Water Use

1. (1) Public water supplies, (2) agricultural uses, (3) rural uses, (4) self-supplied industry.

2. The 1975 national average water use was 168 gpcd. Of this, 4 percent or 68 gpcd was used domestically.

3. 280 gpcd (See *Basic Science Concepts and Applications,* Mathematics Section, Per Capita Water Use.)

4. (1) domestic, (2) commercial, (3) industrial, (4) public.

5. (a) 156 gpcd (See *Basic Science Concepts and Applications,* Mathematics Section, Domestic Water Use Based On Household Fixture Rate.) (b) Lawn watering (i.e. 69 gpcd or 44 percent of total water used.)

6. (a) 1.75 mgd at 8 PM, (b) 0.20 mgd at 2 and 3 AM.

7. b and d

8. 159,000 gpd (See *Basic Science Concepts and Applications,* Mathematics Section, Water Use per Unit of Industrial Product Produced.)

9. Peak day demand is the greatest amount of water used for any one day in the year.

10. Factors affecting water use, with their effects, are:

    Time of day and day of the week—use varies among communities, but tends to stay the same during a season for a single community.

    Climate and season of the year—more water use in warm dry areas and in areas experiencing severe winters.

Type of community—less water use in residential communities than in highly industrial or commercial ones. Larger per person use in low population density residential areas than in high population density areas.

Household water use: increased use of appliances is causing water use per person to increase.

Water pressure—more water use with higher pressure.

Metering—more water use with water-cooled air conditioners and with humidifiers.

Water quality—less water use with certain types of poor quality waters.

Sewers—more water use when sewer systems are in use.

Conditions of the system—leaks in distribution system and overflowing storage tanks waste large amounts of water.

## Module 3   Developing the Water Supply

1. Excavated impoundments, reservoirs, and natural lakes and ponds.

2. Intake structure.

3. Evaporation, seepage, silting, and maintaining good raw water quality.

4. Static water level is the water level in a well when no water is being taken from the aquifer through the well.

5. True.

6. (a) 20 ft, (b) 350 ft, (c) 9 gpm per ft of drawdown (See *Basic Science Concepts and Applications,* Mathematics Section, Well Problems.)

7. If contaminants come within the cone of depression they will probably be drawn into the well and contaminate the water.

8. (1) Piezometric surface, water table or static water level, (2) Drawdown curve or cone of depression, (3) Well casing, (4) Gravel pack, (5) Well screen, (6) Pumping water level, (7) Grout seal, (8) Concrete well slab—sloped to drain, (9) Sanitary seal, (10) Radius of influence, (11) Pump, (12) Discharge.

# Module 4   The Transmission of Water

1. Surface diversion, submerged intakes, floating or moveable intakes, pump intakes, and infiltration galleries. (See intake structure section in text for descriptions.)

2. Aqueducts are usually very large, long, high volume water transmission lines. Types include flumes, tunnels, open channels and surface conduits.

3. (a) (1) Asbestos-Cement Pipe (ACP), (2) Gray Cast Iron Pipe (CIP), (3) Reinforced Concrete Pipe (RCP), (4) Plastic Pipe (PP), (5) Steel Pipe (SP), (6) Ductile Iron Pipe (DIP)
   (b) Refer to "Pipelines and Couplings" section in text for advantages and disadvantages.

4. (1) Bell and spigot, (2) Mechanical, (3) Push-on, (4) Flanged, (5) Ball and socket, (6) Threaded, (7) Sleeve, (8) Solvent weld, (9) Welded steel (Refer to "Pipelines and Couplings" section in text for descriptions.)

5. (1) Stop flow, (2) Regulate flow, (3) Prevent backflow, (4) Drain the line, (5) Bleed off air, (6) Take in air, (7) Control water hammer, (8) Isolate sections of line.

6. True, because the valve disc offers little resistance to flow when turned sideways (opened).

7. To allow the water to be diverted into different flow paths.

8. To relieve water hammer pressure which can damage the transmission line.

9. Air and vacuum relief valve. Releases entrapped air when line is full or being filled; allows air to enter when line is being emptied to prevent pipe collapse.

10. Usually a simple gate valve: used to drain that section of pipe and used to blow-off accumulated sediment.

11. Slamming or closing quickly causes water hammer which can damage the pipes, the valves, and the pumping equipment. Always close valves slowly.

12. Air binding could reduce transmission capacity. Worse, draining the line could cause pipe collapse. Routinely maintain and clean air and vacuum relief valves and make sure they are operating *before* you drain the line.

13. (a) 5, (b) 9, (c) 8, (d) 2, (e) 3, (f) 6, (g) 4 & 1, (h) 7 (See *Basic Science Concepts and Applications,* Hydraulic Section, Pumping Problems—Pump Heads.)

14. 72.7 whp (See *Basic Science Concepts and Applications,* Hydraulics Section, Pumping Problems—Horsepower and Efficiency.)

15. Total head = 26 ft, Pump efficiency = 84 percent, Horsepower = 35 bhp (See *Basic Science Concepts and Applications,* Hydraulics Section, Pumping Problems—Reading Pump Curves.)

16. 65.9 motor horsepower (See *Basic Science Concepts and Applications,* Hydraulics Section, Pumping Problems—Horsepower and Efficiency.)

17. $1.24 per hour (See *Basic Science Concepts and Applications,* Hydraulics Section, Pumping Problems—Pumping Costs.)

18. Centrifugal.

19. Single-stage pump.

20. (1) It is compact, so relatively small space is needed, (2) It is inexpensive to purchase, (3) It is simple to operate, (4) It is simple to maintain and repair.

21. b, c, d

22. Radial-flow and axial-flow.

23. (1) Deep-well installation where space is limited, (2) Booster pumps installed along transmission and distribution systems.

24. Without a check valve, the water in the discharge line will flow back through the pump and perhaps cause damage to the pump and motor. In preventing this backflow, check valves save the cost of repumping.

25. To meter chemical feed rates and flow rates of an individual home service connection.

26. The statement is not necessarily true. V-notch weirs come in a variety of different notch angles including 30, 45, 60, and 90 degrees.

27. Suppressed rectangular weir.

28. So that the depth measurement is not affected by the sloping surface of the water approaching the weir.

29. 7.93 mgd

30. 1.   There are no vertical walls or obstructions across the flow stream to silt-up or collect debris.
    2.   The high rate of flow through a flume helps keep it clean.

31. 1.01 mgd

32. As the name implies, the basic principle is that it measures flows based on *differences in pressure.*

33. 10 cfs (See *Basic Science Concepts and Applications,* Hydraulics Section, Flow Rate Problems—Flow Measuring Devices.)